The Magic of the Scottish

The Magic of the
Scottish Islands

TERRY MARSH
AND JON SPARKS

D&C
David and Charles

A DAVID & CHARLES BOOK
Copyright © David & Charles Limited 2002, 2008

David & Charles is an F+W Publications Inc. company
4700 East Galbraith Road
Cincinnati, OH 45236

First published in 2002
First UK paperback edition published in 2008

Text copyright © Terry Marsh 2002, 2008
Photograph and captions copyright © Jon Sparks 2002, 2008

A catalogue record for this book is available from the British
Library.

ISBN-13: 978-0-7153-1151-6 hardback
ISBN-10: 0-7153-1151-4 hardback

ISBN-13: 978-0-7153-2856-9 paperback
ISBN-10: 0-7153-2856-5 paperback

Printed in China by SNP Leefung Pte Ltd
for David & Charles
Brunel House, Newton Abbot, Devon

Editor: Sue Viccars
Book design: Les Dominey

Visit our website at www.davidandcharles.co.uk

David & Charles books are available from all good bookshops;
alternatively you can contact our Orderline on 0870 9908222
or write to us at FREEPOST EX2 110, D&C Direct, Newton
Abbot, TQ12 4ZZ (no stamp required UK only); US customers
call 800-289-0963 and Canadian customers call 800-840-5220.

Page 1: Evening, Yell Sound and Ness of Sound, from
below Hill of Clothan, Yell, Shetland. Ronas Hill on North
Mainland in cloud. Gruney and Ramna Stacks in the
distance

Pages 2–3: Clouds lit from beneath, over Canna, from
Muck

Right: Traigh Bhalaigh and Bhalaigh (Vallay) island from
Malacleit, North Uist

Front cover: The hills of Skye (Blaven, Glamaig and the
eastern Cuillin) from Raasay, early morning

CONTENTS

INTRODUCTION

There's not a lot going on, until you stop, look and listen. For a mind focused on discovery, there is no part of this island world that doesn't deliver.

How often have we dreamed? – light on calm, crystal waters, freshness in the air, sun warming golden fields and eagles overhead. How often have we dreamed what a magical place this is – this Hebridean isle? But when magic is in the air, how much do we not see?

I set out on this odyssey not knowing how many 'islands' there were around Scotland. Some books mention as many as 800, and I baulked briefly at the thought of exploring them all, then warmed to the idea. But the truth of the matter is that it depends on what you call an 'island'.

In his excellent book *The Scottish Islands* (an essential book for every would-be island traveller), author, artist and yachtsman Hamish Haswell-Smith, after much wringing of hands, arrived at his own definition: 'An island is a piece of land or group of pieces of land which is entirely surrounded by seawater at Lowest Astronomical Tide and to which there is no permanent means of dry access.' He ended up with 163 islands of 40 hectares or more.

But his definition excludes the Isle of Skye, because it is now linked by a bridge, and counts North Uist, Benbecula, South Uist and Eriskay as one island because they are linked by causeways. I felt not a little aggrieved by this pedantry. How could the 'Isle' of Skye no longer be an island?

So, I embarked on a certain amount of soul searching of my own. Did it matter if there was a bridge, or a causeway? No, I thought not – an island is a natural thing and doesn't cease to be a natural thing simply because of man's meddling to suit his convenience. Did it matter if I could walk across to an 'island' at low tide, or if it was permanently linked by a broad sandy tombolo that rarely, if ever, was covered by the sea? No, I thought not – anyway, a tombolo, by definition is 'a bar of sand or gravel connecting an island with another or with the mainland'. So, one end of a tombolo has to be an 'island'. Did size matter? No, I thought not – you don't get diamonds as big as bricks! It was magic we were looking for, not magnitude.

And magic we found.

Hogh Bay, Coll

SHETLAND

Shetlanders know their islands as 'The Old Rock'. And a rocky landscape it is, too, one that, like many of the Scottish islands, either enthrals the visitor or sends them back whence they came, none the wiser.

Bolstered by about 100 smaller islands, the bulk of Shetland comprises the mainland and the three northerly islands – Fetlar, Yell and Unst. They lie almost 100 miles north of the Scottish mainland, and are 70 miles from north to south, a fascinating, sprawling, convoluted archipelago, where land and water intermingle, and as close to Scotland as it is to Norway. Few realise that most of Shetland lies north of the 60th Parallel, which puts it as far north as Hudson Bay in Canada and the southernmost tip of Greenland. This is the land of light nights, of the 'simmer dim', where anyone intent on seeing both a sunset and sunrise will get only a few hours sleep in between.

Not surprisingly, in a place where the Atlantic and the North Sea meet, a coastline that is breathtakingly stunning is something you take as the norm. A confusion of gloups, sounds, wicks, steep-sided 'geos' and deeper, wider 'voes', penetrate the land, and make for a vexed and bewildering geography. You struggle here to get as much as three miles from the sea. Yet in sight of so much water, Shetland is shy of rivers. Here modest burns plough a gentle course through a landscape that is even less tolerant of trees.

UNST

As you stand on the edge of Hermaness Hill on Unst overlooking the skerries of Muckle Flugga and Out Stack (the most northerly bit of Shetland), there's a unique if trivial fascination in the knowledge that there is no one in Britain standing further north than you.

To get here you've driven northwards across the main island of Shetland, making two brief ferry crossings on the way, and finally hoofed it from Burrafirth and across the Burn of Winnaswarta Dale for almost an hour over the peaty mound of the hill itself.

Above: Red campion, Haroldswick, Unst. This island variant has larger and deeper-coloured flowers than its mainland equivalent

Left: Spiggie Loch, South Mainland

Along the way, as you pass through a major breeding ground for great skuas, or bonxies in Shetland-speak, you come under the close scrutiny of parent birds with young to protect. From all directions they swoop in, not always from above as naïvely you anticipate, but cannily circling wide, teasing you to dismiss them as about other business, then, when you're not watching, banking tightly and skimming fast and low – alarmingly fast and impossibly low – over the barren moors before pulling steeply skywards so uncomfortably close you feel the draught of their awesome wings about your face. You soon get the message. Go. Away.

Beyond the bonxie grounds, cliffs plunge to the foam-flecked waves of the Atlantic, between Herma Ness and the knuckle of Tonga. Not dark grey cliffs as geology might suggest, but white cliffs: white with nesting or resting gannets and

Above: Gannet riding an updraught at The Neap, Herma Ness, Unst
Left: A luxurious bus-shelter, near Baltasound, Unst. There is even a visitors' book!
Right: Gannet colony at The Neap, the highest cliff at Herma Ness

patrolled these cliffs, half a hemisphere from home, vainly waiting for a mate.

The impact is awesome, breathtaking, lasting and humbling. You sit. You watch. You listen. You feel privileged. Because that's how to behave when you're in this kind of company. You don't ever want to leave.

FETLAR

On flower-faced, fertile Fetlar, the Garden of Shetland, where whimbrels pipe in nervous unison, a large stone wall known as Finnigert Dyke divides the island in two, north to south, and is probably Fetlar's oldest surviving man-made structure, thought to date from the Bronze Age. Nearby, the Haltadans, a ring of stones circle two central stones. Legend has it that these are a fiddler and his wife playing music for a group of trows, or trolls, caught above ground when the sun came up and turned to stone for their indiscretion.

At Houbie general store, as I wander shelves stacked with everything from nuts and bolts to South East Australian Chardonnay and candles, the postie perches on the counter and plays a tin whistle, quite well too. In the Wick of Tresta dolphins draw the attention of a 'scope' of birders last

Above: The strange 'lunar' landscape of Keen of Hamar National Nature Reserve, Unst
Right: Red-necked phalarope, Loch of Funzie, Fetlar

the guano of a billion birds (or so it seems). Here, acrobatic fulmars (maalies) stiff-wing by and pin you with an Arctic stare; black guillemots (tysties) bob on the swell, red paddles astern; fish-faced puffins (tammy norries) land in a flustered flurry at your feet before going subterranean with lunch for the kids. For years a black-browed albatross

Scraada make sea music, crashing waves bullying through collapsed caves onto washed rocks, drawing back, sighing, surging in again. Chambered cairns litter the landscape, black-backed gulls (swaabies) line Loch of Houlland like spectators at a football match, opportunistic gulls survey the salmon cages out in Yell Sound, waiting – which in Shetland is a fine way to pass the time.

Perch at Eshaness lighthouse and feast on the cliffs and headlands running northwards to Shetland's oldest rocks at Uyea; accompany this ragged frontline in the timeless battle between land and sea at least until Ronas Voe shepherds you inland. Such a walk is a pilgrimage among birds, tenacious, cliff-loving plants and song-filled moorland acres of sky. From the chambered cairn on Ronas Hill all Shetland lies at your feet; northwards the landscape flows, literally, for North Roe is as much water as land, a pageant of Nature's design where every step is an adventure and every adventure a step in the right direction.

Above left: Peat diggings, overlooking Basta Voe, Yell
Below: Archaeological excavations at Old Scatness, South Mainland

seen red-necked-phalarope-spotting along the shores of Loch of Funzie. Back at Oddsta, ferry-waiting, sea wrack became an otter became a rippling display of consummate fishing skills – dive, surface, fish in mouth, hold in paws, eat, throw away remains, if any. Simple.

YELL

From Gutcher on Yell, the road rises across the Hill of Bixsetter, down to Sellafirth and round fingery Basta Voe, largest of Yell's voes. Every rise across this squat island reveals distant views of Yell, of even squatter Hascosay, of Out Skerries and Whalsay, of Eshaness and Shetland's Everest, Ronas Hill, round, soft-shapen and a mere 27,500 feet lower than the real thing, but who's counting?

Blanket peat covers most of the island, making of Yell's face a plain canvas on which it is hard to paint memories. It is one island among many, and, for many, little more than a stepping stone: even the tourist literature calls it the 'gateway to the northern isles'. Such action as there is happens at Mid Yell, the main centre of the local population, which possesses one of the finest natural harbours in Britain, a sheltered refuge that defies all winds.

NORTH MAINLAND

The north-western part of the Shetland mainland is known as Northmavine. It is, without question, scenically breathtaking, a spectacular Atlantic coastline with no comparison. All but a handful of lochs are hallmarked by red-throated diver pairs that hunker down against the wind. The Holes of

CENTRAL MAINLAND

Here more contrasts await: limestone is betrayed by bright patches of green amid the duller brown of harder underlying rocks. Glacier-shaped voes probe the land for weaknesses, tombolos constrain wannabe islands umbilically, and storm-tossed islands offer a million excuses to wander. One impression is of a constant battle, land against raging tempest; another is one of compromise, a working in harmony to produce a landscape of Gothic design and transcendental beauty. In this Nature is successful.

Scalloway, sheltering below the Hill of Berry, was Shetland's capital in the seventeenth century, picturesque and leafy, and protected from Atlantic gales by the bridge-linked islands of Trondra and the Burras. During World War II it was a secret Norwegian base. Here stand the remains of

Opposite: Voe from Hillside, Central Mainland

Left: Clickimin Broch, Lerwick, Mainland

Above: 'The Lodberrie', Lerwick, Mainland

Above: Commercial Street, Lerwick, Mainland

Right: Replica Viking longship in Lerwick harbour

Opposite: Dinghy sailing, Lerwick small boat harbour

Scalloway Castle, built in 1600 by the despotic Earl of Orkney, Patrick Stewart, illegitimate son of James V of Scotland, using forced labour, but occupied for less than a century.

Across the island, Lerwick is a town that grew from an embryonic seventeenth-century settlement developed to meet the needs of Dutch herring *busses* that gathered each year in Bressay Sound. So vigorous was the trade between islanders and the 'Hollanders' that troops were garrisoned at what became Fort Charlotte to defend the British interest in the far north. The narrow shoreline lane, overlooked by tightly packed buildings, which today is stone-flagged Commercial Street, was and is the main drag of Lerwick's social and business life.

Left: Oil platform under repair, Breiwick, Lerwick
Opposite: Scalloway Castle and harbour, Mainland.
Beyond are Green Holm, Papa and Oxna

SOUTH MAINLAND

For most visitors their first sight of Shetland is Sumburgh, standing proud above the sea swell. At night Sumburgh's lighthouse marks the spot, erected, like its sibling on Muckle Flugga at the far end of Shetland, by the monopolistic lighthouse-building Stevenson family, in this case Robert, the grandfather of Robert Louis Stevenson (Muckle Flugga was built by the novelist's father, Thomas).

Robert Stevenson accompanied Sir Walter Scott to Shetland in 1814, on a cruise that produced Scott's novel *The Pirate*, and seemingly survived Sumburgh Roost, scene of many a disaster in the days of sailing ships. Today, Sumburgh is famed neither for sea-defying Fitful Head nor the many-runwayed airport, by-product of World War II, but for Jarlshof. Described as one of the most remarkable archaeological sites ever excavated in

Britain, Jarlshof re-emerged into the world a hundred years ago when violent storms exposed its stonework, rather in the manner of Skara Brae on Orkney. Here, the site is complex and on many levels, from a Stone Age hut, through an Iron Age broch to a Viking village and medieval farmstead – a complete Shetland history lesson in one place.

More recently another sliver of Shetland history appeared when excavations at the tiny Celtic chapel on St Ninian's Isle brought up a hoard of silver ornaments and bowls, believed to date from AD800. This peaceful appendage – peaceful that is when it is not quaking before the full fury of an Atlantic weather system – is linked to Bigton by the finest tombolo in Britain, two sheltered beaches for the price of one.

Further north the broch at Burraland, a ruined, forlorn, history-laden edifice gazes across Mousa Sound to its infinitely more substantial twin on Mousa, the finest example in Britain of an Iron Age broch, and one of more than a hundred built on Shetland alone. Once an eloping lovers' hideaway, today Mousa pulses to the night-time susurrations of nesting storm petrels (alamooties) that return to Mousa under cover of darkness to evade the predatory mobs.

Opposite: St Ninian's Isle and tombolo from the shore at Bigton, South Mainland, with Fitful Head in the distance
Left: Jarlshof, South Mainland: a 'street' in the Viking settlement, looking towards the sixteenth-century Great Hall
Above: A black guillemot (tystie), Mousa

PAPA STOUR

The fertile, volcanic isle of Papa Stour, sometime residence of monks and Duke, later King, Haakon of Norway, sails bravely into St Magnus Bay where it stands full-face to Atlantic moods. Harbours, north, south and east provide shelter for shipping, though the Sound of Papa has a race that can frustrate travel between the island and mainland for days on end.

On Papa Stour's south-west face, Christie's Hole, confined between vertical rock faces at the head of Hirdie Geo, is arguably Britain's finest sea cave, but is challenged for the title by a host of other sea caves and coastal rock formations around the island.

This is no place for those in a hurry, nor one for anyone, like those who came in the 1970s in search of flower-powered nirvana, who sees island life as a prophylactic against the ills of modern society.

BRESSAY AND NOSS

Like Yell, Bressay is treated like a stepping stone, for beyond lies Noss, a passion for lovers of sea-cliff birds. But Bressay has its glad rags, too, fashioned from a bedrock of old red sandstone and shaped into high seacliffs at the south of the island. Elsewhere a fringe of fertile land rings an interior of heather moorland, much favoured by snipe that lace the air with sounds as evocative as that of the golden plover and as certain a signature of wild places.

From the summit of Ander Hill, the Voe of Cullingsburgh, Aith Ness and Aith Voe reach out to mainland Nesting, Whalsay and Out Skerries. The land holds water by the lochful, as if water was in short supply, and the watery confines of

Grimsetter, Brough and Setter become communal leisure centres for shell-shocked refugees from clamorous Noss.

The dinghy across Noss Sound is as close as any non-swimmer will seek voluntarily to come to the half of Shetland that isn't land. Ashore, the island, once used for breeding Durham-mine-destined Shetland ponies, rises from Voe of the Mels across skua-patrolled ground to Noss Head and the sight and sound of 200,000 seabirds for whom the haven of Rumble Wick and the Noup of Noss holds no equal. Guillemots darken the ledges, gannets paint them white, so perfectly white you'll swear it's quartz, and shags adopt

affected poses beyond the reach of spray. They come, they go, they rest, they mate, they overwhelm the place and trellis the air with seabird-speak, trailing calls in their wake, hurling defiance for the smug satisfaction of hurling defiance and hearing it echo off vertical walls like some audio mirror.

Above: Trawler off Bressay
Right: Sea cliffs at Noup of Noss

WHALSAY AND OUT SKERRIES

Man has inhabited low-lying Whalsay for over 4,000 years; almost 2,000 of his stone tools were found inside the Neolithic Beenie House. Certain it is that early man found the fishing to his liking, a trait that has continued ever since. At a mere five miles by two, there is little of Whalsay that isn't coastal, and though the fertile pastures might well support a farming economy, it is to the sea that its people turn for their livelihood.

For centuries the salt-fish trade of Whalsay was managed by German merchants, and every year ships from Europe brought in cloth, iron tools, alcohol and luxury goods. But not all trade was legal. Tradition points to a smugglers' tunnel from the harbour at Symbister to the cellars of the laird's house, the Auld Haa.

Out Skerries, meaning the 'eastern' islands, of which Housay, Bruray and Grunay are the stars, are Shetland's most easterly extreme, and possess the most perfect of natural harbours: Böd Voe, which remains calm in even the most violent of storms. Such a haven was crucial to the island's *haaf* fishing industry. Today, the car ferry bounces out from Lerwick, seemingly hell-bent on an argument with forbidding cliffs. But closer scrutiny opens up a gap in the blockade beyond which lies the calm amid the storm.

The Skerries coastline is low but delightfully rugged, teeming with birdlife and geographically fascinating; in summer sea pinks carpet the cliff tops. Some of the place-names – almost all of them like music to the ears – are a link, like Tammy Tyrie's Hidey Hole, with the days when press gangs ran wild and carried off every able-bodied man they could find to serve in Britain's war against the American colonies, the Dutch, the

French and the Spanish. Elsewhere in Britain, press gangs were an occasional menace, but throughout Shetland the innate prowess of island men and boys with boats and oars fitted them perfectly for a life of misery and hard labour on the seas of war.

FOULA

There is no guarantee with Foula (or for that matter, Fair Isle) that the brief outward flight will be retraced successfully the same day, or the next, or the next. Nor can the ferry service be relied on if an easterly is sending great waves crashing into Ham Voe. Plans for Foula have to be flexible, and outlooks philosophic. This idiosyncrasy is typical of a place that still holds to the twelve-days-adrift Julian calendar, which the rest of the world gave up in 1753, and a people that still spoke Norn, a form of Old Norse much resembling Faeroese, until the end of the nineteenth century.

At 1,370 feet, The Sneug is as high as you can get on Foula, though it seems much higher – it is, for good measure, the second highest point in the whole of Shetland. From the landing strip a ridge rises to Hamnafield and across Brustins to The Sneug, from whence it skitters to a dizzying halt at The Kame, the most dramatic precipice on Foula and second only to Conachair on St Kilda for the accolade of 'highest cliff in Britain'. To describe The Kame as awesome is a masterpiece of understatement.

Opposite: The north end of Foula from Soberlie Hill, with Gaada Stack prominent

Right: Marsh marigolds at Ham, Foula

From this abrupt ending of Foula-world you descend Soberlie Hill north-eastward, intoxicated by the headiness of The Kame, which rears up behind you as you go. But if that isn't assault enough on your sensibilities, then Gaada Stack, Foula's showpiece, awaits. Its pillars tower 130 feet or more over the north coast, with stacks, steep-sided geos and storm beach named, simply, Da Stanes. One of the stacks at the north end is known as Da Broch on which an old stone wall had been built, but the arch collapsed in 1965.

Bonxies and Arctic skuas compete fiercely for breeding space on Foula, with kittiwakes and Arctic terns, but there was a time when, thanks largely to Victorian egg collectors, the population of great skuas was reduced almost to extinction on the island. Now they are back in force, and a force to be reckoned with. The diminutive Leach's petrel is a bird rarely found in numbers elsewhere among the islands, and, like Foula's own breed of big-footed field mouse and 'hard-back' sheep, a touch of the unusual among an island uniquely unusual.

Left: The narrow cleft of Sneck o'da Smallie, on Foula's west coast

Opposite: The Kame, Foula. Figures on top give some idea of the scale (about 1,200 feet high)

Opposite: Mountaineering sheep, Da Nort Bank, Foula

Above: Foula – a close approach by a bonxie (great skua)

FAIR ISLE

With an optimism that in less life-threatening circumstances I would have found touching, Eddie the Pilot plunged the Islander towards the badly ploughed field that was Fair Isle's airstrip.

On impact it proved to be no more than an unevenly graded track across which our seven-seater Reliant Robin with wings hurtled, ill-at-ease, towards an alarmingly large flock of whirling birds. Thankfully, on this trip, the aircraft

Above: The lower, more fertile, southern half of Fair Isle
Opposite: The road to Fair Isle's airstrip runs through the nesting grounds of Arctic terns, which give pedestrians a lively welcome

was also the Air Ambulance, a comforting thought. The relief of being on the ground was soon dispelled when I realised those skuas and terns, both of the Arctic variety, were a gauntlet-run that promised all the fun of a major body piercing session.

Lying equidistant from Shetland and Orkney, Fair Isle is Britain's most isolated inhabited island, though with the evacuation of St Kilda still in their minds, for the few remaining Fair Isle people of the 1940s it must have been a close call between leave and stay. Only one man's vision – of the island as Britain's premier ornithological site – saved the day, and for George Waterston's optimism throughout his years as a German prisoner of war we must all be thankful.

The Norse settlers named the island Fridarey – the island of peace – though it was often used as a look-out and for sending fire signals to and from Shetland. Inevitably it has been a useful landmark for shipping as well as migrant birds, but in mist it has also proven far from helpful, with a least a hundred known shipwrecks about its shores, including the Spanish Armada ship the *Gran Grifon*, which was wrecked here in 1588. Only in the twentieth century did the island acquire a safe harbour, and even in these times it is safer to haul the mailboat out of the water in winter.

The cliffs of Fair Isle, impressive as they are, are not those of Foula. Bu Ness gives a flavour, and tilted Sheep Rock (an iffy place for shepherds in a gale, if ever there was one) more so. But it is along the west coast, where the teeth of the Atlantic have bitten into Fair Isle's fleshy flank, that the elemental drama of Fair Isle unfolds. Here the tougher bits of Fair Isle still defy the sea, and lead northwards to the highest point, Ward Hill, from which the whole island is surveyed. On a good day, Shetland, Orkney and Foula are on display; on a not-so-good day be thankful to see your feet, and even more thankful to be standing on them.

Fair Isle is magically fair; a resilient place with resilient people. The weather comes and the weather goes, but the subtle magic of Fair Isle remains.

Opposite: The north coast of Fair Isle from near North Light, looking towards Stacks of Scroo
Below: View from Malcolm's Head across the south end of Fair Isle, with the main settlement and wind turbines – Sheep Rock is beyond

Left: Big seas around Yesnaby Castle stack, Mainland

ORKNEY

For many people, even as recently as the nineteenth century, Orkney lay beyond the boundaries of civilisation, islands so remote that visitors rarely went there. We know better today, but in 330BC or thereabouts, a Greek traveller from Marseilles claimed he could see the edge of the world from Orkney. In reality, Orkney proved for thousands of years to be very much at the centre of things rather than on the edge. It has been at the crossroads of history in Northern Europe for over 7,000 years, and for the Norse earls who later settled here, the islands were an important staging post in their trade with the Hebrides and Ireland.

The islands have been inhabited since the last Ice Age, and the initial impression today is that Orkney rests beneath a mantle of peace and tranquillity. But there is a bustling industry about Orkney, too, from tourism and oil production to whisky distilling and Orcadian crafts, an energetic cultural life and a staggering wealth of archaeological sites, even a thriving monastic community. All these fragments combine to create Orkney magic, a mosaic that is distinctly unique and precious.

But what binds the fragments together, more than the pervading warmth of its people, more than the remarkable wealth of prehistory – which brought World Heritage Status to the heart of Neolithic Orkney – more than the profundity of its wildlife, more than any feeling of distant independence, is an amazing light. It is a luminescence that marks the slightest nuance in the land, silhouettes dramatic headlands, broods sullenly among the standing stones, makes of the ordinary a bold declamation, and illuminates Orkney's treasures in a way only Nature knows how. The distinction is hard to explain, easier to see for oneself. But paint a picture of Orkney and seven-tenths of it will be sky, one-fifth sea, and the land but a sliver somewhere between the two. It possesses vitality, inspiration and wonder. And when setting sun and sea get their act together, the experience is beyond belief, an enchantment that lasts a lifetime.

Yet the essence of Orkney isn't distilled from its prehistory any more than it can be found by aimlessly wandering the streets of Stromness or the coasts of the remote islands. With patience and years of experience a determined observer might home in on the nub of the matter, as Orcadian writer Eric Linklater did when penning his 1965 book *Orkney and Shetland*. In it he writes of the local newspaper – *The Orcadian* – which 'pays scant attention to news from London or foreign parts, but never fails to report, in a very well-informed column, the seasonal activities of the innumerable birds that

MAINLAND

Focal point of Mainland activity is Kirkwall, built on a narrow neck of land separating East and West Mainland. The town was founded in the eleventh century, but achieved greater notoriety a hundred years later with the building of the magnificent Cathedral of St Magnus, founded by Rognvald III, nephew of Magnus, in 1137. It was made a royal burgh by James III in 1486. The town centre is a compact array of well-ordered, narrow streets of wall-to-wall flagstone, and sandstone dwellings robust enough to toss aside the evil moods of Orkney weather, many of which charge furiously up from Scapa Flow, little more than a mile to the south.

But Kirkwall doesn't have it all its own way. The town may be the administrative and business centre of island life, but the seventeenth-century

Above: Kirkwall harbour and lifeboat
Right: Interior of St Magnus Cathedral, Kirkwall
Opposite: St Magnus Cathedral dominates this evening view of Kirkwall

frequent its cliffs... A visiting actress would not escape attention, but a hoopoe or honey-buzzard, a spoonbill or a little stint, would be most assured of a respectful paragraph'.

On one level at least, therein lies what is important to Orkney and its people. At another level, scratch the surface and you start to see a community togetherness, instinctive co-operation and mutual support at work that is rare on mainland Britain.

market town and fishing port of Stromness, a welcome sight for Firth-tossed visitors, plays a key role in the Orkney story. Whalers would often set out from here, and, during the eighteenth and nineteenth centuries, Stromness was used regularly by traders bound for Hudson Bay in Canada. The mile-long main street of Stromness was built well before the days of the motor vehicles and delivery vans that today jostle for passage in its narrow confines. Rising from it, beckoning like impatient children, rise narrow lanes, one named the Khyber Pass. This is clearly seafaring territory where, as Jim Crumley puts it in *Among Islands*, 'the bar conversation crackles in a dozen tongues... and no one much minds whether you wear cocktail dress or wellies or both'. That simple approach also pervades the poetry of George Mackay Brown who lived in Stromness, and wrote eloquently about the land he knew and loved.

The West Mainland hosts a stunning concentration of prehistoric monuments, most dating from the Neolithic or New Stone Age. Skara Brae's ordered sanctity above the golden shores of the Bay of Skaill is certainly the best-preserved Neolithic village in Northern Europe. Its 'discovery' in the winter of 1850, when a storm flayed the grass from the sand dune known as Skara Brae, brought to light this incredible example of Neolithic life, inhabited before the Egyptians had built their pyramids and centuries before the

Left: Interior of the Bishop's Palace, Kirkwall, with St Magnus Cathedral beyond

Opposite: P&O ferry at Stromness with the hills of Hoy behind

construction of Stonehenge. But, for visitors from Kirkwall, getting there means passing Maes Howe, the Stones of Stenness and the Ring of Brodgar, any of which waylay good intentions. The only certainty here is that this compact area held special significance for the Neolithic and Bronze Age people and formed a great centre of power.

Maes Howe, it seems to me, disappoints: its story is unclear and lies ill-at-ease with the interpretations – sacred mound? burial chamber? place of worship? Viking howff? – being forced upon it – perhaps it is best that way, all the richer for the

Left: Passage and house 2, Skara Brae, Mainland

Above: Entrance passage, Maes Howe chambered tomb, Mainland

Opposite: The Ring of Brodgar and Loch of Stenness, Mainland

Opposite: Sun and rain, Loch of Harray, Mainland, with the hills of Hoy behind

Above: Bay of Firth and Wide Firth from the slopes of Wideford Hill, Mainland

sweet breath of mystery. Among the sharp-edged Stenness Stones, too, mystery prevails as the mind tries to fashion some meaning from their evident order. But among the monoliths of the circle of Brodgar, though no greater enlightenment is bestowed, a sense of purpose seems to flow. And it would take a barren soul to patrol its embrace and not fall beneath its spell or the weight of unanswered questions. In mist, the stones loom above you like proprietorial bouncers, peering from the shroud of antiquity at each interloper.

At the north-westernmost tip of the mainland lies Brough Head, a tidal island bearing the remains of a Norse settlement and a medieval chapel. Nearby stand the ruins of Earl Stewart's Palace, built about 1574 by Earl Robert Stewart, father of Earl Patrick and half-brother to Mary, Queen of Scots. This far-flung corner was once

the seat of power in Orkney, in the days of Earl Thorfinn, known as Thorfinn the Mighty and claimed by some to be one and the same as Macbeth, who acquired the Scottish throne around AD1040. The co-identity is almost certainly fictional, but the alliance between a king of Scotland and a Viking earl who held, as well as Orkney itself, Shetland, the Hebrides and more, demonstrates Orkney's proper place in the history of Scotland and the significance of the humble ruins at Birsay.

Along the north Mainland coast, the enchanted isle and sometime monastic Eynhallow, protected by Eynhallow Sound's tide races and hidden rocks, rides off shorelines peppered with Iron Age settlements – at least four on the mainland and another six on adjacent Rousay. Such a gathering of communities suggests that this was the centre of Iron Age activity in Orkney, in much the same way as the concentration around Maes Howe and Brodgar locates the heart of the island's Neolithic world. The Broch of Gurness is by far the most outstanding relic here, perched bravely on the stubby point of Aiker Ness.

The history of East Mainland is much more modern; the region's virtues lying in its tranquil beauty, its coastline and wildlife. As with West Mainland, the southern border is Scapa Flow, that 60 square-mile expanse of wartime notoriety, a

Left: Victoria Street, Stromness, Mainland
Opposite: Landscape near Maes Howe, Mainland, looking over Loch of Harray. Both the Ring of Brodgar (far right) and the Stones of Stenness (right of centre) are visible, while Ward Hill on Hoy appears at far left

Opposite: Breakers and spray, towards Marwick Head from Birsay, Mainland

Above: Mine Howe, Mainland, a unique and enigmatic Iron Age 'temple'

Right: Broch of Gurness, Mainland: a house interior with the main broch behind

natural harbour protected by the islands of Hoy, Flotta and South Ronaldsay. In 1912, Scapa Flow was identified as the best site for a naval base capable of accommodating Britain's Home and Atlantic fleets in the event of war. When the war was over, the German fleet was brought to Scapa Flow until a decision was made about its future. The German officers, however, had other ideas,

and scuttled their ships where they lay, unwittingly providing prosperity for those then engaged on salvage work and, more than fifty years on, an attraction for tourists.

When war broke out again, up to 100,000 personnel were based in Orkney. Here, in 1939, barely a month into World War II, a German submarine slipped among the British fleet and

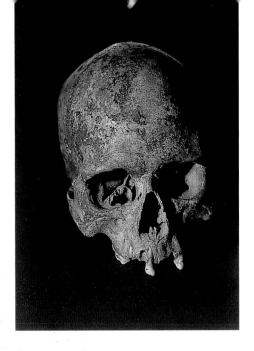

Above: Rainbow over the Italian Chapel on Lamb Holm
Above left: A 5,000-year old male skull from Tomb of the Eagles, South Ronaldsay
Left: Interior of the Italian Chapel, Lamb Holm
Opposite: Sunset over Flotta oil terminal and the hills of Hoy from Hoxa, South Ronaldsay

torpedoed HMS *Royal Oak* with the loss of 833 lives. Churchill later ordered barriers to be built linking East Mainland with the south-eastern islands. Italian prisoners of war provided the labour force, and found time-out to convert two Nissen huts on Lamb Holm into a chapel, a poignant and pious work restored in the 1960s, and a lasting reminder of the enlightened values of the Orcadian Presbyterians. Now you can drive the easy miles across Lamb Holm, Glimps Holm, Burray and South Ronaldsay to St Mary's Kirk at Burwick. Here moss and lichen add colour, staidness and character to a simple building,

constructed, according to legend by a man, Gallus, saved from certain drowning by vowing to build a church dedicated to the Virgin Mary. His footprints on the back of a porpoise sent to rescue him, which was later turned to stone (which seems a mite ungrateful), were forever imprinted in what has otherwise become known as the Penitent's Stone.

HOY

The café at the Scapa Flow visitor centre in Lyness was welcome shelter from the storm. Above the counter a sign read: 'Special Offer – buy one hamburger for the price of two and get another one completely free'. I passed on that, but the soup was scalding and much needed.

Northwards the road mimics the coastline as far as Hoy village leaving untouched the middle-of-nowhere central mass of the island, a boggy loch-laden, burn-laced wilderness ending tersely in west-facing cliffs. North of Rackwick, a path climbs across the base of Moor Fea to an abrupt end that places the Old Man of Hoy starkly below you. Along the cliffs, squadrons of fulmars give master classes in flying, puffins skitter by in a flurry of wings, great black-backed gulls cruise a predatory flight plan. The Old Man, at 450 feet, and the cliffs of St John's Head, twice that and much more, give ferry-bound arrivals a false impression of Orkney, for the rest of the islands

Opposite: St Margaret's Hope, South Ronaldsay

Right: The Old Man of Hoy and St John's Head, Hoy

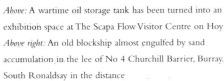

Above: A wartime oil storage tank has been turned into an exhibition space at The Scapa Flow Visitor Centre on Hoy
Above right: An old blockship almost engulfed by sand accumulation in the lee of No 4 Churchill Barrier, Burray. South Ronaldsay in the distance

are low-lying, largely bereft of seacliff drama.

Hoy, its name means 'High Island', is Orkney's second largest island. Between the crofting township of Rackwick and Burra Sound, the road slips through a hill gap below Ward Hill, the island's summit. At the road's high point a path leads to the Dwarfie Stane, thought by Martin Martin to serve some domestic purpose: '…there is … a bed and pillow capable of two persons to lie in … and above … there is a large hole which is supposed was a vent for smoke'. The stone, a huge block of red sandstone, had been the subject of folklore long before Walter Scott immortalised it in *The Pirate*. The reality is imbued with more substance: this is believed to be the only example in the British Isles of a rock-cut tomb of the late Neolithic or early Bronze Age.

ROUSAY, EGILSAY AND WYRE

If prehistory was currency, Rousay would be one of the wealthiest islands in Orkney: there are at least fifteen brochs along the shores of Eynhallow Sound and numerous cairns and burnt mounds. By far the most significant is Midhowe, a chambered cairn that pre-dates Maes Howe, containing twelve burial stalls flanking a central passageway. Nearby stands an Iron Age broch.

At the island's opposite extreme, Faraclett Head yields a Stone Age village – Rinyo – of the Skara Brae fashion, dating to about 3500BC, and

an ancient standing stone, Yetnes-steen (the 'stone of giants'), overlooking the Loch of Scockness, from which it is said to drink at Hogmanay.

In the hilly sanctuary around Muckle Water red-throated divers breed alongside piping golden plover, uneasy tenants in a landscape quartered by keen-eyed hen harriers, merlin, short-eared owls and skuas.

From a distance the twelfth-century tower of St Magnus' Church on Egilsay looks like a grain silo, but such myopic assessments transform on closer inspection into a unique and beautiful structure, once much higher than today. It is one of only a few extant examples of a round-towered Viking church; a nearby stone pillar marks the spot where Earl Magnus was murdered on the orders of Earl Haakon in 1117.

The low island of Wyre is notable for its castle, Cubbie Roo's – more correctly Kolbein Hruga, a twelfth-century chief, described as the most haughty of men, and established in power by Thorfinn the Mighty of Birsay. After his death, legend took over, portraying him as an amiable giant capable of stepping from island to island. In spite of this convenient prowess, he tried to build causeways between the islands, but kept breaking the basket he used for carrying stones, which accounts for the numerous rocks and skerries along this shore.

The beach at Rackwick, Hoy

SHAPINSAY

Shapinsay's green and lush pastures lie far more than a brief ferry ride from Kirkwall. They are found in another world, one far-removed from the dark, amorphous shape pinned to the southern horizon by the spire of St Magnus' Cathedral. Here, amid Shapinsay's sheep and cattle country is the essence of Orkney calm, that quality of exquisite inertia that makes carefree beach-combers of us all.

MV *Shapinsay* throbbed confidently out of Kirkwall's Bay, past Thieves Holm, sometime home of banished thieves and witches, across the flat waters of The Sting and by uninhabited Helliar Holm that throws a sheltering arm around Elwick Bay. Ashore, beyond the ostentatious Dishan Tower, dovecote-cum-saltwater shower, we hovered curiously around Balfour's Victorian ego-piece, its turrets and crenellations a baronial blot on the landscape, made palatable, just, by softening woodland and the redeeming squat grey pleasure of Balfour village, its watermill and gas house (of all things).

Towards the Ness of Ork, Burroughston Broch resists the ocean's quest – but for how long? Safer by far, but maintaining a lonely vigil, lichen-clad Mor Stein rises magnificently from among the ten-acre fields and scattered farms of Housebay and Haughland. Before the witness of such 'meur-steens' ancient assemblies occurred, 'thingmeets', in use from pagan times, places where courting couples came to plight their troth, or the business of the clans debated. The name (closely resembling the Swedish Mora Stone), and position, of the Mor Stein, lend weight to the supposition that this, too, was a 'thing' site.

From this outpost, barely elevated above the heather moor, the view unfolds north and east to Eday, Stronsay and the blue line of Sanday. Westward the vivid green is deepening in the setting sun, a lone heron ponders thoughtfully over a sluggish pond, the ferry beats back across The Sting, and tomorrow is far enough away not to matter.

EDAY

Visitors flying in to Eday's airport may be forgiven a momentary flutter of concern when they hear they are landing at London Airport. Eday's modest airport clings tenaciously to a mere 600 yards of level ground linking the north and south ends of 'Isthmus' Isle (Eithey in the Old Norse).

Loch of Doomy already fills part of that span, and logic suggests that only a matter of time is all that stands between 'Eday North' and 'Eday South', for this is the only break in the island's upland nature.

After Rousay, Eday holds massive appeal for archaeologists. Chambered cairns, the finest on Vinquoy Hill, pepper the hillslopes north-west of Orkney's grandest standing stone, the Stone of Setter. From this weathered monolith long views

Below: The cockpit of a Loganair Britten–Norman Islander over Orkney

Opposite: Looking across Elwick to Balfour village and castle, Shapinsay, with Mainland beyond

open up north to Calf Sound and south over Mill Loch to the Bay of Doomy. But if views were a part of the equation that led to its building, why was it not sited higher?

At Calfsound, Eday's sometime pub is closed, and the ferry to Calf of Eday also a thing of the past. Seventeenth-century Carrick House, where James Gow, the inept and luckless Orkney pirate immortalised by Walter Scott, was captured and held captive, looks out across its bay to the red, seventy-metre high cliffs of Noup Hill.

Eday's simple magic is not for the adventurous, but for seekers after the past, serendipity's companions, and those with time on their hands and little desire to employ them.

STRONSAY AND PAPA STRONSAY

Looking from the air very much as if it is on the brink of being consumed by the encroaching sea, Stronsay's low-lying, treeless landscape and white-sand beaches are today far less frenetic than once they were. In its heyday, Whitehall, the main port, housed fifteen fish-curing stations and as many as 1,500 'fishwives', making this the most active herring port in Orkney by 1900.

To meet the thirsty demands of all this bustle, the island sustained forty pubs, including the original Stronsay Hotel, which had the longest bar in the north of Scotland. During the eighteenth century, kelp farming was important too, and the remains of the industry's kilns can still be found around the shores.

Papa Stronsay was embroiled in the fish-curing business too, in spite of its limited size. The island, which has a plentiful water supply in Mill Loch, was first inhabited by prehistoric man, and much later by monks. In 1046, Papa Stronsay witnessed the murder of Earl Rognvald Brusason by the men of Thorfinn the Mighty, an event which led the islands of Orkney into a long period of relative peace and harmony. Today, Papa Stronsay is again inhabited (indeed owned) by monks, having been purchased at the end of May, 1999, by Transalpine Redemptorists to establish the Golgotha Monastery.

The islands, and adjacent Linga Holm, are a haven for wintering and passage birds, including large seabird colonies and breeding sites for corncrake and quail. Around the island's bays – St Catherine's, Holland, Houseby, Odin and Mill – wildlife abounds, and attracts visitors in search of seasonal rarities. In the days when whales produced both a source of lighting and significant income, the Bay of Holland was a natural trap into which a small fleet of boats would herd the whales to their doom.

Along Mill Bay stands a rock seat known as the Mermaid's Chair, from which one of Orkney's witches, Scota Bess, would cast her spells. It is claimed that any girl who today sits in the Mermaid's Chair will have the power to foresee the future. Further south, Odin Bay has a fine coastline, better endowed with cliffs, admittedly low cliffs, among which shelters the Vat o'Kirbister, a collapsed sea cave, now a gloup, with the finest natural rock arch in Orkney.

Oysterplant near Ayre Sound, Sanday

SANDAY

The elongated island of Sanday is the largest of Orkney's northern isles, and, as its name suggests, its most outstanding features are expansive whitesand beaches. Eighteenth-century maps of the island show a much smaller island than today, suggesting that drifting sand may have joined up a number of small islands.

Kettletoft, once the main landing place for ferries, is now a long way from the pier at Loth, but remains the island's main focal point. Wildlife observation – and a Master's Degree in the Beneficial Qualities of Inertia – are good reasons for visiting Sanday, which has a mellow and relaxing air about it.

From the centre of things at Kettletoft two long arms spread east and west, while Burness makes a less thrusting statement northwards. These peninsulas divide the coastline into a series of attractive bays, much favoured by birdlife, but its low-lying form in the past has been the cause of many a shipwreck, for which, in the days of wooden ships, the islanders were thankful, the island having no peat for fuel!

The humbug-striped lighthouse at Start Point lies on a tidal island to which access is a hazardous (and not recommended) venture. It was first erected in 1802 and in 1806 fitted with the first revolving light in Scotland – another Stevenson invention – but its distinctive colour scheme is a product of the Victorian era, when the lighthouse was rebuilt.

Agriculture has decimated much of the prehistoric traces of Sanday, though remains of the ubiquitous broch can be found dotted about the coastline, and the linked island of Els Ness boasts numerous cairns and a splendid chambered cairn – Quoyness – the largest excavated in Orkney.

To the north, Burness, as well as being a base for the breeding of Angora rabbits, is the site, at Scar, of the Saville Stone, an enormous erratic block of gneiss said to have been thrown from Eday by a witch vexed by a Sanday man with whom her daughter had eloped. Her aim, too, was erratic, and missed her target. More significantly, here were found the remains of a Viking boat-burial containing the remains of three bodies, a child, a man and woman; the site, centuries on, still yielded the man's sword and scabbard, a quiver of arrows and a number of gaming pieces made from whalebone.

Interior of Quoyness chambered cairn, Sanday. There has been some modern 'improvement', including a skylight

WESTRAY AND PAPA WESTRAY

Much of the detail of the Viking occupation of Orkney, which began shortly after AD800, comes from tales embodied in *The Orkneyinga Sagas*, which paint a colourful picture of Westray and its families playing significant roles in the political life of the islands generally during the early twelfth century.

Westray is and always was an independent world, self-reliant and happy to be remote: Westray islanders are said to have been the only inhabitants of Orkney to have supported the Jacobite cause after Culloden. The island today is one of the most productive farming areas in Orkney and the landscape shows this, though the west coast has some fine cliffs, many with caves. Typically, inland from the sea-coast cliffs and headlands a rich maritime heath develops supporting a good range of plantlife, including the Orkney-endemic *Primula scotica*.

Pierowall is the main village, cupped in the embrace of its bay. A short way inland, just north of the Loch of Burness the angular walls of Noltland Castle jar against the lie of the land. This sixteenth-century oddity was built for Gilbert Balfour, who held office under Mary, Queen of Scots, was implicated in the murder of Cardinal Beaton, involved in the murder of Darnley, and later accused of treason. Perhaps Balfour saw Westray as his last refuge. Certainly the building is incongruously well defended, and built on a grand scale, combining the austerity of a military regime with a touch of elegance.

The red sandstone cliffs of Noup Head extend for almost five miles and form part of an important breeding area for seabirds, second only to St Kilda: guillemots, fulmars, razorbills, kittiwakes and

Above: Preparing crabs at Westray Processors Ltd in Pierowall, Westray

Right: Boats in traditional nousts, Bay of Pierowall, with Fitty Hill behind

Page 60: Big seas at Noup Head, Westray

Page 61: Head of the main stairs, Noltland Castle, Westray

puffins fill the air, though the puffins tend to prefer the squat sea stack, the Castle o' Burrian, near the bay of Rackwick. This precipitous stack was the site of an early Christian hermitage.

Linked to Westray by the shortest scheduled air flight in the world (under two minutes), Papa Westray, the 'Priest Island' or Papay as it is known throughout the North Isles, is home of the oldest house in northern Europe, at the Knap of Howar. The settlement has been dated to about 3700BC, and comprises two solidly built stone houses.

Papay is one of the more remote of the Orkney islands, and its tiny four miles by one encompass scenery that ranges from impressive cliffs, at Mull Head, to wide, sandy bays and farmland. North Hill is the island's highest point, and from it the land slopes down to Mull Head and an awesome tide race known as The Bore, where the North Sea and the Atlantic meet in turmoil.

Much of the northern part of the island is a nature reserve for birds; ironic then that Papay was the place where the last surviving Great Auk, a large flightless bird, is said to have been shot.

NORTH RONALDSAY

Isolation succours individuality, and so it is not surprising to find low-lying North Ronaldsay a place where old traditions prevail, Orcadian names predominate and the custom of communal sheep grazing on the seashore practised. It is because of these unique and primitive sheep that a six-foot-high wall, Sheep Dyke, circles the entire island, confining the sheep to the shore, though they enjoy dispensation and a change of diet at lambing time. They are all that remain of an ancient breed of Orcadian sheep; now they exist nowhere else (save for a small flock kept off the

island as a contingency against oil-spill disasters).

North Ronaldsay has been occupied since prehistoric times. Loch Gretchen in the southwest corner of the island has a twelve-foot-high standing stone, the Stan Stane, around which the islanders used to dance on New Year's eve. On South Bay, where the waters of North Ronaldsay Firth lap a sandy beach, the remains of another ancient settlement provoke similarities with Skara Brae, and the Knap of Howar on Westray. Further south still, at Strom Ness, the Broch of Burrian is

the focal point of an extensive Iron Age settlement, occupied well into the time of the Picts.

Traditionally in Orkney, communal farms and smallholdings had defining boundary walls. For the most part these have disappeared, but on North Ronaldsay this is not so, and there remain two impressive attempts at apportioning the island between three brothers. Thought to date from 1000BC, Matches Dyke and Muckle Gersty are enormous earthen dykes that virtually span the island.

The new lighthouse at Sinsoss Point, lit in 1854, is the tallest land-based lighthouse in Britain, rising to over 100 feet. The old lighthouse at Dennis Head, however, is a character of much greater distinction, a fixed light beacon not unlike a chessman in design, but which proved inadequate and was replaced by the light at Start Point on Sanday in 1806.

With such detachment from the main island group, North Ronaldsay, like Fair Isle to the north, is an ideal place for passage migrant birds. On the rocks of Seal Skerry a colony of cormorants host visiting common and grey seals, abundant in these waters, which they share with porpoises and Risso's and white-beaked dolphins.

There is an aura of calm and friendliness about North Ronaldsay, a small island beneath a huge sky.

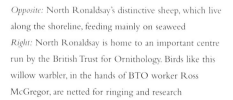

Opposite: North Ronaldsay's distinctive sheep, which live along the shoreline, feeding mainly on seaweed
Right: North Ronaldsay is home to an important centre run by the British Trust for Ornithology. Birds like this willow warbler, in the hands of BTO worker Ross McGregor, are netted for ringing and research

Left: Big seas at Yesnaby, Mainland. A rainbow forms in the spray of a waterfall blown back by high winds

Above: Visitors to the Ring of Brodgar, Mainland, with Loch Harray behind

Right: The lonely grave of Betty Corrigall, Hoy, Orkney

WESTERN ISLES

Athwart Atlantic fury, the Outer Hebrides or Western Isles are a chain of islands, reefs and skerries, often referred to as the Long Island, as if they were one continuous island. They reach for over 130 miles north to south, and have a distinctive culture that has the Gaelic language and traditions at its roots.

In addition to the main components, the Western Isles include numerous smaller islands, the largest and most important of which lie between Harris and North Uist, and Barra and South Uist. And there are yet more: in addition to the main chain several groups of even more remote islands roam the seas, access to which is, for most, little more than a dream – the Shiants, the Monachs, Flannan Isles, North Rona, the St Kilda group and Rockall, the most far-flung outpost of Britain, almost 200 miles west of St Kilda.

To visit any of these islands needs considerably more ingenuity than the ability to read a ferry timetable, and a good measure of patience.

Looking across Traigh Uuige (Uig Sands), Lewis, towards the rugged wilderness which extends into North Harris

These are lonely, remote places, much frequented by wildlife for which they provide safe and peaceful havens. It was not always so. In times past many were inhabited, and the waters around them frequented by trawlers from Fleetwood in Lancashire that had sailed northwards to fish the Atlantic depths. But for many the light of life has gone forever.

LEWIS AND HARRIS

Although thought of as two islands, Lewis and Harris are one, but with markedly differing characters: Lewis, to the north, is low-lying and blanketed with peat moors and lochans; Harris, smaller in area, is rocky, mountainous and blessed with fertile machair along its western coasts.

The mountain barrier, plainly evident as the Calmac ferry furrows Tarbert-bound from Uig on Skye, nevertheless comes as a shock to the system. The crude, natural, raw power is awesome and majestic at the same time, arousing a traveller's urge to explore, to lose oneself in the folds of Nature.

Quite where the dividing line between Harris and Lewis occurs is a moot point. Maps tell you one thing, road signs another. Yet division

Meavaig and Langadale. Beyond this mountain checkpoint the land defies road building. Land and water meet in equal proportions – reflecting lochans creating the illusion that the light is rising from the ground – reaching northwards to the ancient stones of Callanish, giants turned to stone by St Kieran, they say, for their refusal to accept Christ.

Cruciform in shape, the stones of Callanish stand alongside the Ring of Brodgar on Orkney as the most important monolithic circles in Scotland. More than 4,000 years old, the stones have an alignment that seems to relate to periods of the moon. Overlooking Loch Ceann Hulavig lesser stone circles ride the moor, like the rest, their true purpose lost in obscurity and there only to be guessed at. As if expecting an answer, we wander into their midst, we speculate, we call upon the Gods of Earth's Mysteries for answers, we try to divine a meaning, we stand bemused, awe-struck, reverently touching the stones, feeling the hard fabric of their lichenous coats, their rough texture, their sun-blessed warmth. And then we leave, none the wiser, but suitably impressed, as if by association with pious company we have been purified, a confessional that has relieved us of our sins. It's an unsettling sensation – awe, humility, respect, lack of understanding, and a strong conviction that to be in this company we must have been promoted.

North of Callanish, gazing down on Loch Carloway and sheltered from Atlantic tempers by the island of Great Bernera, stands one of the

there is, marked by history and discernibly differing cultures that, in spite of a common language, have distinctly different dialects.

With a good deal of involuntary casting about, the road linking Tarbert with the island's only town, Stornoway, for a while parallels the shores of Loch Seaforth. Between there and the more easterly Loch Erisort lies a rugged wilderness known simply as 'Park'. Here red deer stride freely across a landscape largely inaccessible except by boat across Loch Seaforth or from the minor road that runs south-east from Loch

Strandavat to Seaforth Head. No great heights lie here, but among these hidden glens you touch the rainbows, smell the acres of starlit skies, breathe the pure air of cleansing and wrap yourself in the warm comfort of solitude.

West of Loch Seaforth, with a marked disinclination to penetrate the wilderness, a road replicates the coastline to beautiful Hushinish Bay and the small island of Scarp, testing ground for rocket mail. North of this, the ancient Forest of Harris is hacked into great glens to challenge the adventurous – Cravadale, Ulladale, Chliostair,

Above left: The harbour at Stornoway, Lewis

Opposite: The standing stones at Callanish, Lewis

best-preserved brochs in Britain, second only to Mousa in Shetland. Parts of its galleries and stairways are still intact and display a prehistoric craftsmanship that can only be marvelled at. Dun Carloway is about half the age of the Callanish Stones, and like those elsewhere in Scotland, built at the start of Christian times. They appear as fortresses against seaborne attack, but are little more than secure homesteads to which their occupants would resort in times of trouble rather like a tortoise retreating into its carapace.

What is not so obvious about these structures is that they are a clear statement that at the time of Christ a settled, organised agricultural community existed in this extreme part of the Britain, with the skill to build a structure that would last for 2,000 years and more, and to think logically about a design that would best serve their purpose.

Those same construction skills were employed further along the seaboard, at Arnol, where a traditional black-house, a *tigh dubh*, demonstrates how well these 'simple' dwellings were matched to the climate, setting walls six feet thick beneath a strong roof of thatch and turf to maintain a dry and warm interior whatever vexations the ocean at the door had in mind. The Black House at Arnol was occupied well into the 1960s.

The west coast road continues beyond Loch Mor Barvas, passed Scotland's tallest standing stone at Ballantrushal, up to the Port of Ness from where Lewis men sail each year to gather young gannets – guga – which are a great delicacy, but not thought so by everyone. At the Butt of Lewis the land ends and plunges to crumbling stacks to which the sea clings grimly, forever in combat.

Much further south, a mere 700 yards is all that binds North and South Harris. In the square at Tarbert, the Screen Machine, a mobile cinema that tours the highlands and islands throughout the year playing to packed audiences, unfolds its capacious wings in readiness for tonight's showing of *Chicken Run*. Local children are having an impromptu 'disco' on the stage, self-consciously dancing to Hear'Say.

Eastwards an undulating road reaches to the new bridge that tethers Scalpay, temporary refuge of fleeing Bonnie Prince Charlie, to mainland Harris. Neat, bright cottages cram every nook and press hard against the rocky hillsides, raised vegetable beds tussle with the unremitting rocks beneath to provide rootage for their crops.

Southwards, beyond the ferry terminal, a minor road – the so-called Golden Road – skirts the impenetrable heart of South Harris, taking the only conceivable dry route through an exquisite mosaic of lochans and moor. The name strikes a note of irony, given, as sometime editor of the *Stornoway Gazette* James Shaw Grant remarks, 'to mark the anguish of a remote and grudging authority over the cost of providing a road' seventy-five years after a local minister had reported that the absence of a road and bridges was an 'unspeakable hardship'. Less ironically, this truly is a 'golden' road, for it plunders a magnificent scenic heritage, as rare and deserving of protection as any in Scotland.

Left: The pier, North Harbour, Scalpay
Below: The nine-hole golf course at Scarasta, Harris, has a setting no championship venue can match
Right: Cliffs and lighthouse, Butt of Lewis
Page 74: Abhainn Ulladail, Harris, looking out over Frith Mhorsgail, with the overhanging profile of Sron Ulladail on the right
Page 75: Rodel, Harris, looking over St Clement's Church to the Sound of Harris and North Uist

On the Atlantic side, the main road, heading for Leverburgh, skirts the vast expanse of Traigh Losgaintir beyond which the beaches of Taransay decorate its soft green form as sweetly as grace notes adorn a melody.

Leverburgh is a testimony to the patriarchal good intentions of Viscount Leverhulme, who wanted to see economic development in Harris. Before his arrival, this part of Harris had been economically stagnant, much as it had been left after the cold draught of the Clearances had swept through. The village that now bears his name, called Obbe, was small and peaceful. Suddenly roads were built, houses appeared as if overnight, water supplies were put in and piers constructed. Just as suddenly it stopped, one fateful afternoon in May 1925, when Lord Leverhulme died.

Opposite: Evening, high tide, Traigh Losgaintir, Harris, looking to Taransay. The calmness is deceptive: when the wind drops, the midges come out
Above right: Machair at Huisinis, Harris, looking to Scarp
Right: Mrs Campbell, handloom weaver of Harris Tweed at Plocrapol, near Tarbert, Harris

earfund

recent Christmas appeal,
tor Cuthbert works with
ie has seen an amazing
s like yours.

of water and sanitation
e villagers' faces. There's a
lants to improve crop
upport people living with

NORTH UIST, BENBECULA, SOUTH UIST

With skill I can only begin to imagine, and risk I care not to think about, the ferry jiggled its way through the rocky menaces that dot the Sound of Harris and somehow found the appropriate bit of North Uist to bump into, skirting en route the low mound of Berneray, favoured by royalty in chilling out mode, it has been said, and linked to North Uist by a causeway, opened in 1998.

On a modest hill, Ben Langass, overlooking a *chiaroscuro* fantasia of dark and light, land and lochan, I entered Barpa Langass chambered cairn on hands and knees, of necessity. At Carinish, the Temple of the Trinity (Teampull na Trionaid) is a sad relic, steeped in history, and an important archaeological site, once a seat of learning similar to Iona.

Opposite: The stone circle, Finn's People, looking over Loch Langais to Eabhal, North Uist

Above right: Scolpaig Tower, North Uist, with Loch Scolpaig behind and Eilean Hasgeir on the horizon

Right: Field with corn marigolds, Balranald National Nature Reserve, North Uist

Below: Post box by Scolpaig Tower, North Uist

Opposite: Looking north-east from Rueval, Benbecula, over Grimsay to North Uist, with Eaval on the right and the mountains of Harris in the far distance. The string of causeways linking the islands can be seen left of centre

Above: Balivanich and the Monach Isles from Rueval, Benbecula

Swapping a leisurely three miles for a hasty 200 yards enables a brief tour of Grimsay, a mere comma between the phrases of North Uist and Benbecula, and a stepping stone, too, between the Catholic south and the Protestant north. It is, for all its brevity, a most beautiful island, the centre of North Uist's fishing industry.

Benbecula is inordinately flat, a low-lying island as much jigsaw lochans as land. Rueval, its only hill, creeps marginally above 390 feet, but

rarely does the land exceed 100. The difficulty is in thinking of it as anything other than a stepping stone between the two Uists. What action there is, is happening in Balivanich. Here, fuelled by the army base, the voices crackle not only in Gaelic, but in Geordie, Cockney and Cornish, brought here by military duty.

Presenting the rugged heights of Hecla and Beinn Mhor to the east and fertile machair to the west, South Uist is scenically the finest of the

southern group of islands that comprise the Long Island. Down the whole length of the west coast, where the main road fashions a life-giving spinal cord along the edge of the hills, springtime brings raptures of colour and heavy-scented flowers set against great runs of gleaming white-sand beaches. The range of contrasts is considerable, from lime-rich machair to acidic moorland peat, from sea-washed beach to windswept crags. Great sea lochs penetrate the eastern coast and in the south-

ernmost and largest of these the eponymous village of Lochboisdale has grown.

Further north, the birthplace of Flora Macdonald lies in sad ruins down a side road. Here it was that she came and met the Young Pretender, a chance encounter that inscribed her name forever in the history books. A few miles away, Loch Druidibeg is a nature reserve and a wildfowl sanctuary *par excellence*, an important breeding ground for greylag geese and the

Above: South Ford, Benbecula, looking towards Hecla and Beinn Mhor on South Uist
Opposite: Evening, Sound of Barra, from near Smeircleit, South Uist

greatest concentration of wading birds in Britain, with the sort of supporting cast from gadwall to jack snipe, peregrine to red-necked phalarope that would have birdwatchers scurrying in circles for days on end.

ERISKAY

Between South Uist and Barra, Eriskay is an innocuous island, or so it seems. But the island has a place in history for two completely distinct reasons. At Coilleag a'Phrionnsa, on its west coast, on 23 July 1745, less than nine months before defeat at Culloden, Prince Charles Edward Stuart, Bonnie Prince Charlie, landed from the French ship *Du Teillay*; it was the first time he stood on Scottish soil.

Two hundred years later, at the height of the Battle of the Atlantic when German U-boats were exacting a terrible toll on merchant shipping, New York-bound steamship *The Politician*, endeavouring to evade detection by slipping through the notorious Sound of Eriskay, foundered near the tiny island of Calvay. On board, among motorcycle parts, silks and plumbing fittings, were more than 250,000 bottles of whisky. Once the crew were saved, the islanders turned their attentions to this bounty, steadily

Opposite: Hecla, Beinn Corradail and Beinn Mhor from Loch Druidibeg, South Uist

Below: Coilleag a' Phrionnsa (the Prince's Beach), Eriskay

removing the bulk of the cargo over the next few weeks, a process that all accounts describe as increasingly inebriate. When Customs and Excise officers appeared on the scene, every available orifice and container, from hot-water bottles to rabbit holes, concealed the precious liquor. Sir Compton Mackenzie's hilarious account of the tale in *Whisky Galore* makes no mention of the people that went to prison for illegal possession, or the fact that for years afterwards, even to the present day, bottles of whisky still turn up in the most unlikely of places.

BARRA/VATERSAY

Blue waters surround the island of Barra, as they do all the Scottish islands. But those around Barra are the bluest of the blues. Every shade of blue; every temperament; every infinitesimal gradation of blue imaginable from the borders with green to the fringes of indigo.

In the shop in Castlebay, my request for a newspaper was greeted with the information that they wouldn't be in until after 3.30 – provided the aircraft could land. Out on Tràigh Mhór, which serves as the islands' airfield, we watched it do so, and take off again for Benbecula, sand and spray arcing in its wake. The wait for its arrival, its turnaround and subsequent take-off consumed a good hour, car tailgate up and kettle permanently on the boil. Across the Sound of Barra, Eriskay shimmered lazily in the heat: Gighay, Hellisay and Fuday seemed to hover on silver sand, a shapely fin, a basking shark, broke the surface beyond Orosay, lingered and sank back beneath the ocean calm. Less than a mile away, in stark contrast, Atlantic breakers battered the western coast with great, elongated rolling pins of tumbling waves.

Above: Interior of St Michael's church, Eriskay, with its unusual altar incorporating the bow of a life boat from HMS *Hermes*
Above right: Madonna and child, slopes of Heaval, Barra
Below right: A BA/Loganair flight landing at Tràigh Mhór, Barra's 'airfield'
Opposite: Castlebay, Kisimul Castle and Heaval, Barra

North of the 'airfield', the tiny cemetery of Cille-bharraidh is the last resting place of the MacNeils, clan chiefs of the island, with their stronghold in Kisimul Castle in Castle Bay. In the church to which the cemetery belongs was found the only Hebridean example of a Norse sculpted stone, a replica now in its place. Among the tombstones lies Sir Compton Mackenzie, who made Barra his home.

Linked by a causeway constructed in 1990, Vatersay is the southernmost inhabited island of the Western Isles. Its hill, Heishival Mór, conceals from Barra's view the pinched waist of sandy dunes, a mere 400 yards wide, sandwiched between Vatersay Bay and Bàgh Siar. The island's economy today is based on sheep and cattle, and the causeway, like pedestrian crossings in so many urban settings, only constructed after a fatality. In Vatersay's case, it was Bernie the bull, a prize bull who was drowned as he swam the Sound of Vatersay, as island cattle had done for centuries before, bound for market, or, in Bernie's case, services to be rendered.

ST KILDA

To arrive by air or by sea? that was the question.

Nature writer, Jim Crumley, has no qualms: 'St Kilda should be greeted from the sea, from a berserk deck burrowing blindly west into oceans of winds, deserts of waves the size of dunes.'

And so it was. Eventually. Ignominiously, the first attempt floundered when the Atlantic, in a moment of petulance, decided to toss my intended conveyance – MV *Elektron* – back onto the rocks of Village Bay where she had off-loaded supplies.

The court of St Kilda is not a conventional ferry ride, but it must surely be one of the loca-

tions most sought after by those who explore wild and inaccessible places, a far-flung domain of islands, surrounded by implacable, uncompromising seas. From them the islands rise as gaunt, dramatic rock statements, architectural sculptings on an unimaginable scale, appearing unexpectedly through the grey-green mist like silent portents of doom, or growing ever larger across an ocean swell, still glowering, as if by force of habit, even on a calm day.

The air is frantic with converging streams of seabirds – gannets, fulmars, razorbills, guillemots, puffins, and Leach's petrels, too – Atlantic seals jacuzzi in such foam-filled backwaters as Glen

Opposite: Rippled sand and reflections on the west coast of Barra

Above: The village, Hirta, St Kilda

Bay and its Tunnel allow, bonxies busy themselves about your head, and the sound of St Kilda's own breed of wren penetrates the moan of the wind and constant sound of the sea.

This distant realm – Hirta, Boreray, Soay, Dun and isolated stacks – is collectively known as St Kilda, though the name has come to be assigned principally to Hirta. There never was a saint Kilda, and the name probably derives from the Norse, *tobar kelda*, a touch of tautology since both words, one Gaelic the other Norse, mean 'well' or 'spring'.

The bird's-eye view of Village Bay and the natural breakwater of Dun from the steep slopes of Conachair stands among the finest panoramas among all the Scottish islands. Between embracing arms, a semicircle of white waves delineate the shore, in the foreground the row of squat houses, half-derelict dykes and scattered dry-stone 'cleits', primitive refrigerators hunched across the landscape like waymark cairns. Behind the village tough brown grasses snap at the heels of the once-cultivated land and make a frame for the picture that is Village Bay. Beyond the land rises, on the right to Oiseval and Conachair, mast-topped Mullach Mór and the defiant, angular thrust of Mullach Bi. Suddenly you realise that the island's shapely hills have no backs: they rise smoothly, then stop, as if hacked in two, and plummet a thousand feet to the sea.

It seems so small, more perilous, more intimidating, this dot on the map that focuses all its concentration on that sheltered bay and its grey-stone village.

And it takes the village to remind us that these sea-girt sentinels were once home to a breed of people whose fortitude we cannot begin to imagine. For over 3,500 years, the islands were inhabited; in the summer of 2000, archaeologists discovered the remains of an Iron Age building that had lain undisturbed for almost 2,300 years. Here they recovered fragments of pottery and water-carrying vessels, some dated to 300 years before Christ.

Most of today's *des res's* date from Victorian times, the original village having been demolished in 1830 when 'new' black-houses were built. A hundred years later, on 29 August 1930, the few remaining St Kildans, no longer able to sustain their lifestyle, left the islands. Ironically, everything they needed to maintain their existence is now on Hirta, but, for a population that knew nothing of water supplies, sewerage treatment, money, or, for that matter, trees, their home today would be as alien as the mainland locations to which they were moved.

No longer do the stone walls echo to the laughter of children, the sounds of raised voices, the soft murmuring of human intimacy, the sound of men and women labouring to eke out an existence... nor the chatter of gossip.

Isolation provided the amber in which their lives were preserved; emigration shattered the illusion.

Opposite: Ruined cleit (and wheatear), Mullach Sgar, Hirta, St Kilda, looking towards Dun

Above: Cliffs east of Conachair, Hirta, St Kilda, looking to Boreray, with Stac Lee on its left partially hiding Stac an Armin

THE LOST ISLANDS

In the last 200 years, numerous Outer Hebridean islands have, like the St Kilda group, become uninhabited, losing forever an often unique culture and distinctive way of life developed in isolation over many centuries. Some were depopulated during the Clearances, others by the inexorable effects of disease, loss of men at sea, emigrations or, simply, resignation in the face of the difficulties of existing in such remote places.

North Rona is the most northerly of the Outer Hebrides ever to have been regularly inhabited. It lies forty-four miles north of the Butt of Lewis, more distant even than St Kilda, and in the sixteenth century had a flourishing community. Towards the end of the seventeenth century the island was devastated by rats and a raid by a passing ship, and finally deserted in 1844.

The Flannan Islands were never permanently inhabited, so far as records show, though a chapel dedicated to St Flannan is known to have existed here a thousand years ago. Approaching Christmas time in 1900, however, the island was the hub of a great mystery when its three light-house keepers vanished in mysterious circumstances leaving a ready meal untouched on the table. Around this time there is known to have been one of the most severe storms ever recorded, and the lighthouse keepers are thought to have been caught unawares by a follow up tidal wave of the same magnitude that separated the Monach Islands from North Uist many years earlier.

The Shiant Islands, off the east coast of Harris, saw the last family leave in 1901, though at its greatest the population never rose above sixteen.

Possibly owing its name to a Pictish noble-man called Tarainus, mentioned by Adomnan in his *Life of St Columba*, the island of **Taransay** sits low off the coast of Harris, uninhabited and vir-tually unknown until a television programme in 2000 brought it unnecessary notoriety. Flourishing since the sixteenth century, at one time the island accommodated a thriving com-munity, but by 1971 its population had reduced to one family of three people, and then the island remained empty, save for sheep and wildlife, for over twenty years.

But had you been there on the night of 2 November 2000, all was unusually still, the wind so feeble it made no impact on the windmill, the sky was clear and the stars bright in their heaven. Across the machair drifted the sound of human voices joined in harmony as a motley choir of fourteen rehearsed 'Silent Night' for the forth-coming Christmas festivities. When silence returned the only sound was that of waves lapping indolently on the seaweed-garlanded shore. Overhead, shooting stars streaked across the night sky, a blaze of glory all-too-soon dispelled. Stillness returned and the island slept.

Eight miles offshore, **the Monach Islands**,

known also by the Norse name, Heisker, breast the Atlantic waves. These five small islands, little more than raised sandy beaches, were once linked at low tide to North Uist. But in the sixteenth century a huge tidal wave swept away the sand banks, isolating the Monachs and their inhabitants forever. The last families left the Monachs in 1943 after the closure of the lighthouse.

One evening in September 2000, as I watched from the shores of North Uist, the sun slipped through the clouds of an artist's sky towards the horizon, momentarily sending a shaft of sunlight, aimed, it seemed, for the Monachs. For a few magical seconds – or perhaps it was an hour – the islands sailed on a silver pond, dark, low shapes barely breaching the far Atlantic rim: the dorsal fin of a killer whale broke surface off the island's edge, then that of its calf. Then they were gone, and with them a moment of privilege.

For the people of **Mingulay**, their home was Eilean mo Chridhe, Isle of my Heart. As Alisdair Alpin MacGregor explains in *Islands by the Score*, Mingulay was their world 'sometimes halcyon, oftener tempestuous. After Barra itself, theirs was the largest of the archipelago long known as the Barra Isles'. Southwards lay Berneray, to the north Pabbay, Sandray and Vatersay.

The west coast of the island, where, many years ago the young and agile men of Mingulay snared guillemots and razorbills, is a dramatic plunge to the sea, reminiscent of those on St Kilda.

Berneray is the true end of this remarkable 'Long Island', the cliffs of Skate Point taking the leading edge of the Atlantic gales, often a force so mighty that, according to the records of lighthouse keepers, in 1836 a forty-two ton rock was moved almost two metres by the force of a storm.

In the seventeenth century, the waters around the islands were rich in fish, but, as Martin Martin noted, the islanders too canny to go fishing while the MacNeil chief or his steward were on the island, for fear they would raise their rents.

Opposite: Old graveyard near Sand Hill, Berneray (North Uist), looking to the hills of Skye

Above: Barra – Kismul Castle and MV *Isle of Mull*

SKYE AND THE SMALL ISLES

SKYE

The most scenically spectacular and diverse of all the Scottish islands, of Skye one thing is certain: you either assume an instant dislike of the place, or you fall unreservedly in love with it, remaining forever entangled in half-coherent enthusiasms for its intangible charm.

For most visitors, there are no half measures: with Skye it's all or nothing. No other of Scotland's islands excites such feelings; it is as if Skye has an air of magic, a numinence, denied to other islands, a veil of mystery that incites strong and enduring emotions. This malaise is known as 'Skye fever'; it is all part of Skye's magic. There is no cure.

Among the Hebrides, Skye is second in size only to Lewis, having an area of 535 square miles and a coastline so indented that no part of the island is more than five miles from the sea. From Rubha Hunish in the north to Point of Sleat, the distance is forty-nine miles, and if you travelled cross-country in a straight line from east to west,

Opposite: Evening, looking to Eigg from near Godag on Muck, with Blaven (Skye) at left
Right: The lighthouse at Neist Point, Skye

apart from being close to exhaustion and sick of the sight of peat bog, you would cover twenty-seven miles.

Skye spreads its wings into The Minch like some great bird about to alight on its prey. Reason enough to call it An t-Eilean Sgiathanach, the Winged Isle, but Skye is also known as Eilean a'Cheo, the Isle of Mist, a term of romantic rather than negative allusion. When the tourism spin-doctors brought their marketing tentacles to Skye's door, euphemisms erupted like a bad rash – The Isle of Enchantment, the Isle of Mystery, the Isle of Fantasy.

It was 'over the sea' to Skye that Bonnie Prince Charlie, aided by Flora MacDonald, sailed from Uist during his flight to France following the disaster at Culloden in 1746. And where, in 1773, Dr Samuel Johnson came with Scottish biographer and diarist James Boswell, only to find the decay inspired by the Whigs' and Tories' determination to destroy the clan system and annul the Gaelic language, already well advanced. Many believe that Skye and the whole of north-west Scotland were, during the eighteenth century, intentionally devastated, and in the nineteenth, left without much support or interest from the corridors of power in far-off London.

Of today's visitors to Skye, many will be heading for the Cuillin – which offer the finest mountain walking in Britain – and the bar of the

Left: Moonrise over Loch Bracadale from the Ardroag road, Skye
Opposite: Summit of Dun Caan, Raasay looking over Loch na Meilich and Loch na Mna. The Red Hills of Skye (looking very blue) are in the distance, with the Cuillin in cloud on the right

Opposite: Loch Scavaig and misty Cuillin Hills from Elgol
Above: Flower meadow, Glen Brittle, Skye, with the Cuillin Hills (Sgurr na Banachdich and Sgurr Dearg) behind

Sligachan Hotel, though not always in that order, for there are days when the Cuillin are lost in mist. Others come to fish, paint, study the wildlife, or simply to seek their roots. But whatever the reason, most also come because Skye has especial, if inexplicable, appeal and the power of recall, bringing visitors back to explore again its winding trails, its glens, its coasts, and its history of dark and bloody intrigue.

Like most of Scotland's islands, Skye is espe-

Left: The Fairy Pools, Coire na Creiche, below the
Cuillin Hills, Skye
Above: Towards Waterstein Head from Neist Point

Right: Bernie Carter climbing Cioch Slab Corner
(Difficult), the Cuillin Hills, Skye

Left: The strange rock formations of the Quiraing, Skye:
looking up to the Needle

Above: Trotternish from the Quiraing

Opposite: The pinnacles of the Storr, Skye: the Cathedral
and the Old Man from the Sanctuary

cially well seen from the sea, whether it be across the dramatic cliffs of Trotternish or Waterstein, the awesome spectacle of the Cuillin from the heart of Loch Coruisk, or, indeed, from its neighbouring islands. In the 1930s, after a long train journey from the south, to Inverness, and another exhilarating train ride from Inverness through Strath Carron to Kyle of Lochalsh, visitors to Skye had to join the queue for a steamer, first the *Glencoe*, then the *Fusilier*, and later the *Lochnevis*, ships of David MacBrayne's embryonic fleet, which

would take them serenely up the sound between Skye and Raasay to Portree Bay. There was a special magic in that, something that today most can only imagine.

Since the opening of the bridge at Kyle, only two ferry links with the mainland remain. The best of these, and it probably always was the best, is the crossing from Mallaig to Armadale, followed by the delightfully lush introduction to Skye that Sleat provides. Road 'improvements' help speed you along, but there's enough of the old, single track

Above: The Skye Bridge from near Kyle of Lochalsh
Opposite: Fishing boats at Portree
Right: The Glenelg ferry at Kylerhea

Left: Low tide, Portree, Skye
Below: Roag Island, Harlosh and the Cuillin from Greep
Right: Marsco, Sgurr nan Gillean, Am Bhasteir and Bruach na Frithe from Loch nan Eilean

road remaining – for the present – to slow visitors down long enough to admire this frequently passed-through but rarely visited corner of Skye. And beyond, the Island (with a capital 'I') awaits.

The magic of Skye, since we're dealing in magic, comes ultimately from the interplay of sea, land and sky, in combinations so varied and frequently changing that no two visits are ever alike. But for many, part of the magic flows from re-acquaintance with agreeable memories: a sudden

Above: Evening, the Sound of Raasay and Trotternish from Rubh' an Uillt Dharaich, Skye
Right: Ardtreck Point, Wiay and Idrigill Point from Boust Hill, Skye, with South Uist on the horizon

Above: Stone mermaid and Raasay House
Opposite: Sunrise on the Red Hills and Cuillin of Skye, from Raasay

shaft of sunlight on a loch, perhaps, or the first sight of the Cuillin's jagged edge, the joy of seeing an otter or a heron at the water's edge, or an eagle overhead. Others find among its quiet ways a relaxation and escape, the clichéd 'getting away from it all', a chance to unwind and refresh. But quite why Skye should do this better than anywhere else cannot be explained. Perhaps it doesn't need to be.

Large enough to forget occasionally that this is an island, Skye is itself surrounded by a collection of smaller islands, most uninhabited, though not necessarily *never* inhabited – Ascrib, Flodigarry, Staffin, Holm Island, Scalpay, Pabay, Oronsay, Wiay, Tarner, Harlosh, Roag, Isay and Mingay. But, hazily-remembered, there was delight in lazing with friends in the scented grass

of Oronsay, hauling the dinghy onto the rocky shore of Flodigarry, and tracing the setting sun's progress as Sleat's shadow climbed the light on Isle Ornsay. It is of memories such as these that Skye's magic is woven; all you need is to believe in magic!

RAASAY

Busty mermaids are perhaps the least expected of welcomes among the Scottish islands, but one such decorates the Battery on Raasay's shore, gazing across to the rugged profile of Skye.

If it is true that Skye exists to provide shelter for Raasay, then in return Raasay provides a perfect antidote to high-speed Skye. On Raasay, sufficiently distanced from 'mainland' Skye to provide breathing space, peace and solitude are more readily found, the more so the further north you go.

Long and thin, twelve miles by two (or so), Raasay is composed of gneiss rock, that rises to the distinctive flat top of Dùn Caan. Its summit serves as a stunning viewpoint over the hilly southern part of the island and the comparatively lower northern extremes. Here, on 10 September 1773, in a day that covered twenty-four miles of rugged going, Boswell, having left Dr Johnson for the day, and attended by 'Old Mr. Malcolm M'Leod' and two other gentlemen, danced on the summit, returning later 'not at all fatigued' to take part in the evening ceilidh.

The road running northwards from Brochel to Arnish is one of the most remarkable feats of single-handed engineering. It is a lasting monument to Calum MacLeod, lighthousekeeper and crofter who managed the northernmost croft on Raasay. For nigh on forty years the crofters had

asked for a road to be built, but the authorities declined to provide one. So, Calum decided to build one himself, armed with a four-shilling book on road building. The road consumed the remaining years of the man's life. Sadly, in 1988, while working on one final stretch to his cottage door, he died. Visitors drive the road now, through the tortured, primitive landscape of north Raasay, in a matter of minutes, but greater pleasure surely comes from surveying it on foot, and taking time to marvel at one man's achievement and determination.

Overlooking Clachan Bay, Raasay House was for years the family seat of the MacLeods of Raasay. At the time of the Jacobite uprising in 1745, the Chief of the Raasay MacLeods mustered no less than one hundred men, a surprising, some would say eccentric, act. Many who supported Prince Charles' cause did so because they shared his Catholic faith, but MacLeod of Raasay and his people were Protestant. It was a chivalrous decision, but one for which Raasay paid dearly. In reprisal, and for sheltering the Young Pretender on the island for two nights, Government troops burned down their homes, destroyed their boats and livestock, and even, according to one contemporary witness, 'ravished a poor girl that was blind, and most unmercifully lashed with cords two men, one of which soon after died'.

Following the general amnesty of 1747, Malcolm MacLeod returned to his island and rebuilt Raasay House, entertaining Johnson and

Boswell there. Such social niceties did not continue, however. By 1843, the Chief, having struggled through a series of poor, harvest-wrecking summers, the collapse of the kelp industry and a fall in the price of cattle, was facing bankruptcy. He tried the by-now conventional remedy of clearing crofts and introducing sheep, but to no avail, and the island was sold at London auction for 35,000 guineas, with which its former owner set sail for a new life in Tasmania. The evictions and privations of the time were immortalised in verse by Raasay-born poet Sorley MacLean, whose verse encapsulated the heart of his island, of neighbouring Skye and the spirit of the native people of these islands. Raasay House, meanwhile, though in use again, as an outdoor centre and an ideal base for exploring on foot, seems forever to have lost the air of dignity befitting a clan chief's home.

RONA

In the dim and distant days of the sixteenth century, Rona, the rocky, northerly continuation of Raasay, was thickly wooded and the abode of pirates and 'broken men' who hid on the island and raided shipping from Acairseid Mhór, a fine natural harbour, known at the time as Port nan Robaireann (Port of the Robbers). Rona is also known as South Rona, to distinguish it from another Rona, north of Lewis.

During the Clearances, Rona's impoverished ground – Boswell described it as 'of so rocky a soil that it appears to be a pavement' – was occupied by crofters evicted from Raasay. They settled at Doire na Guaile, Acairseid Mhór and Acairseid Thioram, but struggled to eke a living from the harsh landscape. The three settlements, the largest

being Acairseid Thioram, boasted a church and two schools, but today you will find only overgrown remains to tell of those struggles.

Acairseid Mhór is still the principal anchorage, sheltered from view by Eilean Garbh. By the slipway in 1840, a tiny cottage was occupied by the Mackenzie family, at that time the only residents there. In the summer of that year, Kenneth Mackenzie was lost at sea, but his widow, refusing to believe it, kept a light burning in the hope of his safe return. This proved of such great assistance to other vessels entering the harbour that the Navy eventually gave her an award of £20, and the Commissioners of the Northern Lights a small sum to buy lamp oil. Janet Mackenzie kept the light burning for twelve years before she finally gave up, conceded that her husband was dead, and emigrated to Australia with her children.

On the east coast lies Church Cave, a huge cavern with seats and an altar of natural rock. Before the island had its own church (1912), the islanders worshipped there. Even after the church was built, it was the tradition to have babies baptised in the cave: water dripping from the roof conveniently fell into a depression in a stone, and this served as a font. The last service, conducted by Rev Dr James Matheson was held in 1970, attended by thirty members of the church in Portree.

SOAY

Riding in Soay Sound like a ship at anchor, the island is pinched into a dumbbell shape by the narrow inlet of its harbour on the north-west coast and by the bay of Camas nan Gall, where the few islanders live, on the south-east.

This low-lying, seemingly innocuous isle in the shadow of the Skye Cuillin, may not seem a place of technological innovation, but its solar telephone exchange was the first in the world of its kind. And it was here, in the north-west inlet, that Gavin Maxwell (1914–69), writer, traveller and conservationist, best known for his trilogy of books that began with *Ring of Bright Water*, founded the Isle of Soay Shark Fisheries after he bought the island from the MacLeods of Dunvegan in 1946. The dilapidated buildings of Maxwell's enterprise still remain, rusting, forlorn and tumbling down, and full of fisherman's debris.

Owned by the MacLeods since the thirteenth century, by 1832 only one family was living on Soay, crofting. But as the Clearances bit into the population of Skye, more than a hundred dispossessed crofters chose to settle on the island's infertile Torridonian sandstone rather than go to America.

Maxwell's enterprise foundered after only three years, and by 1953, with poor communica-

tions as the principle reason, the islanders petitioned the government to be evacuated. So it was that in June 1953, amid great publicity and to the sound of a piper's lament, the SS *Hebrides* evacuated the crofters to Craignure on Mull, where land had been bought for them by the government, at Java Lodge. All except one family, that of Joseph 'Tex' Geddes, harpooner for Maxwell, left the island. A succession of 'tranquillity seekers' then began to reach the island, staying or leaving as they found or failed to find the quiet life they were searching for, until finally the situation stabilised.

Only a handful of folk live on Soay today. Geddes' house is empty, bracken encroaches on Maxwell's enterprise, and from the bay, above which Solan geese and golden eagles patrol, Oliver Davies fishes for shrimps and whatever else the waters provide for his living.

The north-eastern part of the island, smaller and higher than the south-west, rises to 460 feet in Beinn Bhreac, the highest point on the island. The sporadic woodland on the east side, Doire Mhor (the big oak wood) and a recurring name theme in the rough hilly ground in the south-western, reflects a time when there was far more woodland than is evident today. Only the birch, alder, rowan and oak that grow across the painfully thin waistline of Soay hint at what the landscape may once have looked like.

Left: The Bullough Mausoleum and Harris Bay, Rum
Opposite: The great hall of Kinloch Castle, Rum

THE PARISH OF THE SMALL ISLES

The romantic name 'The Parish of the Small Isles' is given to a cluster of four islands – Rum, Eigg, Muck and Canna – which lie between the Ardnamurchan Peninsula and Skye. Each has unique characteristics, and they are far enough removed from one another to have a magic of their own. Visiting here is like going back in time, in the most complimentary sense, that is. The absence of car ferries, mains electricity, pubs and countless other 'benefits' of mainland society, have helped the islands retain an aura of timelessness.

RUM

There is probably no more unlikely a place than Rum, the largest island of the Small Isles, on which to encounter a Greek temple, but visitors to remote Glen Harris find just that, in the form of the Doric temple-styled Bullough Mausoleum, perched on the edge of rugged cliffs. The Bulloughs were a Lancashire family of textile machinery industrialists, who owned Rum from 1888 to 1957, and the mausoleum, built by Sir George Bullough, contains his tomb, that of his father and his wife.

Little is known about the prehistory of Rum, though it was certainly one of the earliest places in Scotland settled by man. Although once under the control of the Norse and with Norse names commonplace, there is no evidence of Norse settlement. Like so many Hebridean islands, Rum passed to the Scottish Crown following the defeat of King Haakon at Largs.

Today, Rum is owned by Scottish Natural Heritage, and it was here that in the 1970s and 1980s white-tailed eagles were reintroduced in Scotland, having last been seen in Skye in 1916.

Opposite: Askival, Sgurr nan Gillean, Ainshval and Trollaval from Cnapan Breaca, Rum

Above: The south side of Kinloch Castle, Rum

closed to the public – for deer research – has the remains of a once large village and an old burial ground. All three are linked by a roadway, with rough tracks also running from Kinloch, south to Dibidil and on to Papadil.

Rum is renowned for its hill walking, having its own range of Cuillin, less demanding than those on neighbouring Skye, but no less spectacular. The three northernmost peaks – Hallival, Askival (Rum's highest peak) and Trallval – are the remains of a volcano that last erupted in Tertiary times. But Rum, too, is widely appreciated for its plant and wildlife, with almost 2,000 plants and almost 200 species of bird being recorded here.

EIGG

Once owned by a German artist known as Maruma, Eigg became the subject of national attention when the islanders launched a public appeal to raise funds to enable them to buy the island. Like so many of Scotland's islands, the problems of absentee owners dogged Eigg, too, and led to considerable difficulties. The islanders' aims were fulfilled in June 1997, when the island was bought by the Isle of Eigg Heritage Trust, a partnership of the islanders' appeal funds, the Scottish Wildlife Trust and the Highland Council.

Most of the population lives across the shoulder of the island, in Cleadale, though the ferries arrive at Galmisdale, taking their chances with the ocean swell between Eigg and Eilean Chathastail. Mid-way between the two communities, remotely isolated by all but island standards, are the post office and shop, focus of island news and most things social, including a glass of beer, al fresco on a fine day. Eigg has the largest population of the Small Isles and is the centre of parish life.

The ruggedness and sheer scale of Rum is captivating, even from a distance, when it could easily be mistaken for Skye. As you draw closer, passing through the oakwoods south of Morar, the Cuillin of Rum dominate the seascape westwards, towering above the other small isles. Closer still, as you pull into Loch Scresort, tree-clad slopes flank the sound, drawing the eye to the incongruous castle of the Bulloughs.

The island is a rough oval shape, about nine miles from the sandy Kilmory Bay in the north to the rocky headland of Rubha nam Meirleach in the south, and a little less from east to west. Harris Bay, dividing an otherwise unbroken line of cliffs along the south-west coast, boasts little other than the mausoleum, while Kilmory, an area normally

Opposite: On the ridge of The Sgurr, Eigg, looking over Beinn
Tighe to the Cuillin of Rum, with South Uist on the horizon

Above: Kildonan Bay and The Sgurr, Eigg

Kildonnan, just to the north of Galmisdale, is the site of a seventh-century monastery in which St Donan and fifty or so monks lived until in 617 they were murdered by pirates, though some tales accuse a tribe of warrior women who lived on the crannog in Loch nam Ban Móra. The loch and a group of smaller lochans lie in a hollow in the shadow of An Sgurr, the massive conning tower mountain that dominates the island.

An Sgurr is the largest remaining mass of columnar pitchstone lava in Britain, a great look-out, surrounded by brackeny, basaltic moors, and surveying the intimacy of neighbouring islands. In the north of Eigg, the cliffs are sandstone and, at Camus Sgiotaig, have eroded into bizarre shapes. Here, the white quartzite beach is famed for its singing sands, which when dry emit a droning sound and when wet squeak eerily. The sands also boast arguably the finest view in the whole of the Inner Hebrides, that embracing shapely Cuillined Rum nearby.

The far north of the island is low lying, its cliffs set back from the water's edge, a tell-tale sign of a change in sea level since the Ice Age. Offshore, on tiny Eilean Thuilm, can be found the fossilised remains of reptiles and fish. Elsewhere on Eigg, in the nineteenth century the remains of a sea-going dinosaur, a plesiosaur, were discovered.

Eigg is managed today as a wildlife reserve, and its flora and fauna is not dissimilar to the adjacent islands. One curious find, however, is a group of eucalyptus trees that stand out against the indigenous willow, hazel and uncontrollable bracken, an inheritance from the days of mono-culture beef farming.

For such a small place, Eigg has experienced a turbulent past. In one particularly unhappy episode, said to have occurred in 1577, the MacLeods of Harris, in reprisal for ill treatment meted out to a daughter of MacLeod's, married to the captain of Clanranald, suffocated almost 400 MacDonalds, by lighting a fire at the entrance to a cave (still known as 'Massacre Cave') in which they were hiding. Little more than a decade later, however, the island was plundered by Spanish mercenaries from a galleon wrecked in Tobermory Bay, which suggests either that the island was repopulated by MacDonalds following the massacre, or that accounts of the incident are exaggerated, if at all true. There is certainly no reference to the Massacre of Eigg, either in contemporary writing or in government archives. Several MacLeod chiefs have been accused of the deed, though the date puts it in the time of Norman, the Twelfth Chief.

The MacDonalds of Eigg supported the Jacobite cause in 1745, and their leader, John MacDonald, facing the arms of a Government captain, John Ferguson, sent to the island to arrest him, surrendered against the promise that there would be no reprisals against his clansmen. As soon as MacDonald was safely in custody, Ferguson ransacked the island and deported the young men.

During the proscription of the Catholic church following the 1745 rebellion, another cave, close by Massacre Cave, was used for services; this, debatably more authentic than its near neighbour, is known as Cathedral Cave.

Today, Eigg is a peaceful, relaxing place. Its hills and cliffs provide diversions for walkers and climbers, and for birdwatchers, too, for the burrows of An Sgurr shelter large numbers of Manx shearwater, which are often seen drifting in rafts offshore.

MUCK

Low-lying as a submarine about to dive, tiny Muck, the smallest and least well known of the Small Isles, can be strolled around in a day, provided the determined Atlantic winds, which can strand you on the island for days, don't sweep you away. This is an island for relaxing, exploration and quiet contemplation. The main settlement is at Port Mór in the south-east, a scattering of cottages, above which, on the south side of the harbour, stand a few remains of a Bronze Age fort.

Like Eigg, Muck also received the plundering attentions of the Spanish mercenaries, who had been employed by Lachlan MacLean of Mull. Boswell and Johnson passed this way in 1773, and noted that the island was once again prosperous, considerably managed by the laird, fertile and surrounded by waters abounding in fish. Many of the islanders were engaged in the production of kelp, but when the market collapsed, over 150 islanders were deported to Nova Scotia.

The landscape of Muck is formed from layers of basalt that form terraced cliffs and ledges. Dolerite dykes run north-west–south-east culminating in the island's highest point, Beinn Airein, from which the patchwork fields of yellow and green flow away to the rocky line of the coast.

The island is almost completely given over to mixed farming – sheep, cattle, a dairy herd, potatoes, root crops and oats – and is self-sufficient, although the absence of peat means that all fuel, such that can't be gathered from the seashore, has to be imported.

Cathedral Cave, Eigg

Above: Port Mór, Muck

Opposite: Looking towards Rum and Skye from the beach at Gallanach, Muck

Above: The summit of Beinn Airein on Muck, looking to Coll and Tiree

Opposite: The north coast of Canna, looking towards Compass Hill

CANNA AND SANDAY

Historically, Canna belonged to the monastery on Iona, and later fell under Norse control, although the monks continued to cultivate the island, until the Reformation, at least.

The island is divided into two by a neck of land linking Tarbert Bay and Camas Thairbearnais. To the west lies a long, cliff-edged plateau; to the east a gathering of hillocks, burns and hollows culminating in the crags of Compass Hill, so called because of the distorting influence of the rocks on compasses, aided by a vein of iron ore under the seabed to the north. Exposed to southerly gales, Tarbert Bay, which has a fine escarpment, above which are the remains of an ancient nunnery and a ruined fort, is central to the island.

Towards the end of the nineteenth century, Canna was sold to a Glasgow shipowner, and managed sensitively by him to the extent that when his family sought to sell the island after his death, they went to some lengths to find an understanding buyer, in the form of Dr John Lorne Campbell, a farming, historian folklorist. This remarkable man, who came to the island with his American wife, Margaret Fay Shaw, author of *Folksongs and Folklore of South Uist*, spent most of his life preserving the Gaelic culture, and in book-strewn Canna House, his home and the principal house on the island, gathered together the world's largest collection of Celtic language and literature. Today, the island belongs to the National Trust for Scotland.

Canna House is surrounded by woodland and overlooks the adjacent island of Sanday, to which Canna is linked by a bridge. When the bridge was destroyed in a storm, Dr Campbell, a

tenacious man, battled with the district council over its maintenance, and finally produced a copy of the Council minutes, which proved that they had built the bridge in 1905, and so were responsible for rebuilding it. This begrudging concession from the local authority was almost the limit of what Campbell could secure for his islanders: he himself provided the pier, the water, the electricity, and he maintained the roads. And it is to his vision that today we owe the preservation of Canna, one of the most beautiful of the Hebridean islands.

Between Canna and Sanday lies a well-protected harbour, entered by rounding a rock face that over the years has been 'embellished' by graffiti, originally the names of fishing boats and yachts, and a tradition to mark safe arrival in the harbour.

More so than on other Small Isles, many of the tree species have been introduced, notably sycamore, ash, elm, beech, Sitka spruce, Corsican pine, Austrian pine and Japanese larch. Most are found on the hillsides above the harbour, now occasionally patrolled by the once-indigenous white-tailed eagle on a sortie from Rum. Here, too, is A'Chill, the site of the original settlement, and the remnants of St Columba's seventh-century chapel. An Coroghon, to the east, is a prison tower set on an isolated stack in which a clan chief imprisoned his wife to frustrate the attentions of her lover, a Skye MacLeod.

Left: Graffiti, near the pier, Canna
Opposite: The harbour and Sanday from Compass Hill, Canna

MULL AND THE SURROUNDING ISLANDS

MULL

I flatter myself that I saw her before she saw me, but I know that's nonsense. Thankfully I am no mountain hare, brown or blue, for the exposed, rocky slopes of Ben More rising from the saddle with A'Chioch are no place to be when a golden eagle is out for lunch. She slipped effortlessly by, close enough to see the yellow of her eye, as she could see the white of mine. You feel less an intruder when a golden eagle wings in, gives you a companionable nod, and slides on by, undeterred.

I reached Loch na Keal at the foot of the Abhainn na h-Uamha, having taken my fill of Inch Kenneth, Eorsa and shapely Ulva, as the sun dipped out of sight beyond Iona; the others behind me on the hill were not so fortunate, stumbling down the slopes of Gleann na Beinne Fada in failing light.

Opposite: Looking to Ellenabeich village on Seil, from Easdale
Right: The lighthouse on Eilean Musdile, with Lismore behind, from the Mull ferry

A few miles away, at Gruline, stands the mausoleum of Major-General Lachlan Macquarie, son of Ulva, father of Australia, governor of New South Wales.

Located west of Oban and across the Firth of Lorn, Mull is a large island, the third largest Scottish island – so large in fact that you easily forget it is an island – separated from the mainland of Morvern by the Sound of Mull. Its coastline is a convoluted affair, a ragged 300 miles of inlets and lochs, huge chunks approachable only on foot. Roads there are, 150 miles of them, but for many of those miles the impression is one of roads that are an evil necessity, barely on the retention side of superfluous. The roads on Mull do not go overboard, they do not intrude any more than they must, they maintain faith with the contours of the land when they can or cling

Left: Waterfall details, near the fossil tree, Ardmeanach, Mull
Above: Sea pink and basalt, near the fossil tree, Ardmeanach
Opposite: The Carsaig Arches, Mull

deferentially to the edges, fearful of intruding into the heart of this island.

Man has certainly made his impression on Mull. The island has a pedigree of prehistoric remains from standing stones to loch crannogs, churches to chieftain's castles. Duns and forts litter the coastlines, lending weight to the supposition that this was the front line between the kingdom of the Picts and that of Dalriada, kingdom of the Scots. Centuries of Norwegian control ended when the Lord of the Isles gained suzerainty in 1266, from whence flowed a chequered history that brought together the destinies of the MacDonalds, the Campbells, the Mackinnons and the Macleans until James VI/I, determined to eradicate the Gaelic culture, had all the Hebridean chiefs imprisoned on a ship in Salen Bay and transported to Edinburgh.

Opposite: Duart Castle, Mull, from the ferry

Above: Shag, commonly seen off Scotland's islands

Left: Beech woods, near Tavool, Ardmeanach, Mull

Page 134: The fossil tree, Ardmeanach

Page 135: A'Chioch and Ben More from the north, Mull

For those in pursuit of history, the castles of Mull remain – Duart, Aros, Torosay, Moy, Dunara, Glackindaline – in various conditions of repair. But it is not among these that the magic of this isle will be found. For that, patience is needed; that and a good pair of legs and the will to wander, for rarely does Mull come to you.

In the mountain corries, below the island's only Munro, Ben More, and across the twisted Torosay landscape of the south-east, red deer roam, as do feral goats, descendants, no doubt, of beasts turned loose during the shadowy days of the clearances. Not far from the Iona ferry port, otters feast on captive brown trout in the spread of Loch Poit na h-I, as they enjoy crustaceans along the coastline of the Ross of Mull. Sea eagles, offspring of reintroduced stock on Rum, patrol the coastlines, and head a constantly changing cast of birdlife that features all you would expect to see and some you would not.

Above: Minke whale, east of Cairns of Coll, with Ardnamurchan behind, seen on a trip from Tobermory
Right: Otter, Loch na Keal, Mull
Opposite: Morning, Tobermory

Sheltered from all but the worst weather by Calve Island, Tobermory ranks among the loveliest small towns in Scotland, its brightly painted seafront cottages a hallmark famed afar. Beyond the town, a coastal path runs out to the lighthouse at Rubha nan Gall, beyond which lies the curve of Bloody Bay, scene of a great battle between the last Lord of the Isles, John, and his son Angus. On this day, so it is said, the tide ran red ashore, staining the legs and beaks of oystercatchers forever.

Across the north of the isle a twisting road leads agreeably through the hills to Dervaig, beautiful Maclean-built village with its modern forty-seater theatre, and on to Calgary Bay, gazing out to the low, grey islands of Coll and Tiree.

IONA

Seven years before the day of judgement, the sea will rise and sweep over Ireland and Islay; only the isle of St Columba will survive the waves.

Such old prophecies are still currency on Iona, and underpin a numinous faith that sees the island as indestructible. This is a pilgrim's isle, no more, no less. And if saintly associations fail to draw you here, then its geological antiquity must, for few will walk on places much older than Iona. Here, the Lewisian gneiss is among the oldest rock in Britain, the cradle of Nature, if not of Christianity.

A pilgrim's isle indeed, but what kind of pilgrim? As the often-claimed birthplace of Christianity in Scotland, Iona must defer to Whithorn, where the wandering St Ninian

Opposite: Evening, Calgary Bay, Mull
Above left: Cow and calf by the hill road from Dervaig to Torloisk, Mull
Above: Quiet moments in the grounds of Iona Abbey

Above: The Augustinian Nunnery, Iona
Above right: Prayer cross in the abbey church
Opposite: Cloisters, Iona Abbey

worked among a Christian community at least a hundred years before Columba landed on Iona's southern shore. For the tourist, there is nothing, the island is scarcely large enough: ditto the long-distance walker, the backpacker, the adventurer, the seeker of bright lights. Mountaineers will find little to recompense the brief voyage across the Sound of Iona, until they scale down their aspirations and expectations. Then from Dun I's windy top, the spine of the island, its bays of white sand, the ocean richly blue and the distant shape of higher, snow-capped things, on Mull, far-off Skye and the mainland, will offer as much reward to a receptive mind as any much higher peak. All the pleasure without the effort.

Visitors today scarcely notice the village, bound as they are, bent against the wind or caressed by a summer's sun, for the abbey. Long before you reach St Martin's Cross, you pass the Nunnery, a Benedictine foundation built by Ranald of the Isles. In ruins now, its pink granite walls frame gardens of clover and daisies and a contentment far greater than ruins should possess.

The abbey is a place of pilgrimage for everyone, Godly and otherwise, for it is a monument of creation, testament to skill, ingenuity, faith and determination. South of the abbey stands Saint Oran's Chapel, built on Reilig Oran, an ancient burial ground. Here lie the remains of forty-eight crowned kings of Scotland, four kings of Ireland, eight of Norway and one, it is said, of France.

Everyone should see the abbey, walk its cloisters, breathe its pious air, share the tranquillity it offers. And then walk out across the thin waist of the island to the white sands of Camas Cuil an t-Saimh at Shian, and south to Port na Curaich, where Columba put ashore. Explore the rough ground south-west of Dun I, and wander along the northern shore, facing out to the pellets of land that are the Treshnish Isles... and see if you don't become one of Iona's pilgrims.

STAFFA

Uninhabited since 1800, isolated Staffa rose to fame on the wings of serendipity and a bout of bad weather that drove the ship of Sir Joseph Banks, President of the Royal Society, into the Sound of Mull in the summer of 1772. On his return to London, Banks spoke enthusiastically of Staffa, and soon the tourists came, among them Walter Scott, William Wordsworth, Queen Victoria (though she never actually stepped onto the island), Robert Louis Stevenson and Felix Mendelssohn, the composer. Inspired by the experience, Mendelssohn, who visited Staffa in 1829, wrote his *Hebrides Overture*.

Unlikely as you are to experience the island for yourself, it is then that its magic is most tangibly felt. The famed cave is unquestionably a geological wonder, but the simple climb to the island's summit at the south-western edge brings oceans of calm, a spell that easily exceeds all the hype, and a real desire to stay longer, much longer.

Left: Fingal's Cave, Staffa
Opposite: Blackthorn, Ulva, looking to Ardmeanach on Mull and the islands of Inch Kenneth, Geasgill Mor and Iona

TRESHNISH ISLES

Barely a bouncy hour's boat trip from Fhionnport on Mull, or the Ulva ferry, the volcanic Treshnish Islands, of which only the largest, Lunga, is freely accessible, are a fabulous haven for seabirds. During the time of the Viking domination of the Hebrides, the islands were the boundary between the Nordreys, or 'Northern Isles' and the Sudreys.

Southernmost of the group, Bac Mór, the Dutchman's Cap, is a distinctive landmark, a low-lying, green lava ring formed around an old lava cone. Landing on Bac Mór is difficult, but there is evidence, in the form of summer shielings, that the local people regularly managed it in the past.

Below Cruachan, the highest point of Lunga, the remains of black-houses are stark testimony to the former occupation of the island, which ended in 1824, though summer occupation continued for over thirty years more. The noise from the bird colonies is raucous, heavy with guillemot-speak, especially around the separated stack of Dun Cruit, and the air pungent with the smell of layered guano. This notable exception aside, there is a paradoxically fresh and invigorating calm about Lunga, a Site of Special Scientific Interest. Full-beaked puffins land inches from your feet, wink and disappear into burrows; razorbills are more circumspect. From Cruachan's top, a fine view north-east enfolds the skerries and smaller islands of Fladda and the two Cairn na Burghs – Beag and More.

On Cairn na Burgh More a fortification stands on the site of an earlier Norse building. In the thirteenth century, this castle was in the possession of the Lords of Lorn, the MacDougalls, but they lost this and its neighbour, when they were granted to the Lord of the Isles by King David II in 1343. From this splendid vantage point, the MacDougalls held control of the sea-lanes between the Hebrides and Ireland as the Vikings had done before them. The Lord of the Isles in turn placed them in the care of the MacLeans of Duart. One MacLean, heir to the Lochbuie estate and determined to safeguard it, imprisoned the Lochbuie chief on Cairn na Burgh More, his only company an old woman who, against all probability, bore Lochbuie an heir. Enraged, MacLean ordered the child killed, but the old woman had given birth to twins, and handed over only her daughter. The son, Murachadh Gearr, survived to return to Lochbuie to claim the estate.

ULVA AND GOMETRA

For many visitors, the attractive volcanic island of Ulva, a brief ferry ride across the Sound of Ulva, will come as an agreeable surprise. Its lavic terraces, which give the island a distinct, pyramidal profile, are edged, especially at The Castle on the south-east shoreline, with basaltic columns such as are also found on Staffa. There is a lush, summer broadleaf-and-bracken greenness about Ulva that largely conceals a dark, fine-grained igneous rock.

Ulva is the largest of the Mull islands, its small population concentrated in the east, close to the ferry, and its houses surrounded by mixed woodland. Neither the sound of cars (of which there are none), nor that of farm vehicles penetrate the

peaceful 'ambling ambience of an island preserved in aspic', as Mairi Hedderwick described it.

Offshore, to the south, a number of smaller islands gather – Little Colonsay, Inch Kenneth, named after a contemporary of St Columba and, like Eorsa in the mouth of Loch na Keal, once owned by the Prioress of Iona.

When the travelling twosome, Johnson and Boswell, visited Ulva in October 1773, they were advised that 'there was nothing worthy of observation'. The MacQuarries then owned the island, and the visitors were entertained by the Clan Chief. One of the clan, Lachlan, who was a lad of twelve at the time, rose to become a major-general in the British Army, and, between 1810 and 1820, was Governor-General of New South Wales, where he made an impressive contribution to the development of the region, establishing the first Colonial Bank, setting up a school for aboriginal children, and opening an orphanage in Sydney. He is also renowned for his compassionate attitude towards convicts, going to great lengths to integrate released prisoners into the local community. This earned him lasting respect but also unfair criticism, which, in the end, saw him brought back to England.

On Ulva today, there is abundant evidence of the island's former crofting communities, especially alongside the rough track that leads to the neighbouring island of Gometra, a splendid wild walk watched over by the island's red deer. In the first half of the nineteenth century, there were over 600 people living on Ulva, mostly involved in the production of kelp ash, used in making soap and glass. When the potato famine struck, the owner, a solicitor from Stirling, decided that the only solution to the resultant destitution was

to reduce the number of tenants and the conversion of the land to sheep grazing. His methods were nothing less than ruthless, and between 1846 and 1851, he deported over two-thirds of the population, evicting families from their crofts without warning by setting fire to the thatch, denying them the opportunity to salvage their meagre possessions, and seizing their cattle.

At the far end of Ulva, the small island of Gometra, once the home of the renowned Himalayan explorer, Hugh Ruttledge, leader of the 1933 attempt on Everest, is gained by a narrow bridge, beyond which the track continues across the island to Acairseid Mhór, a shallow anchorage adjacent to the tidal Eilean Dioghlum.

Above: House on Inch Kenneth, with the stepped terrain of 'traps' or lava flows on Ulva behind
Opposite: The east end of Ulva, looking to Ben More on Mull

COLL

With few 'Collachs' still living on the island, Coll, in spite of mainstay farming is fast becoming an island economy of holiday homes, lived in by part-time incomers drawn here by a simple splendour and extravagant flora and fauna.

Most of the resident population of around 150 live in or close to Arinagour (the Shieling of the Goats), the only village, and port of call for the Calmac ferry. On entering the village, you pass along a row of neat, whitewashed cottages hunkered down against the elements. They were built around 1800 by the laird of Coll, part of his plan to modernise the island. Small Arinagour may be, but it has well-stocked shops, a post office, a hotel, two churches and a homely and idiosyncratic craft shop.

People have lived on Coll for thousands of years: two standing stones at Totronald, known as Na Sgeulachan, are believed to date back 6,000 years. More recently, people lived in crannogs, man-made islands in the numerous lochans. In much more recent times, the population survived on a staple diet of potatoes, grown in lazybeds, until the Potato Famine of 1846 devastated people's lives and fuelled large-scale emigration.

There is no clear evidence of the derivation of the name, 'Coll', though it certainly appears to be pre-Norse, because it is mentioned in Adomnan's *Life of St Columba*. Nevertheless, Coll was a place of some importance during the Norse

Left: Marsh marigolds, with the two castles at Breachacha behind, Coll

Opposite: The south-west half of Coll from the true (south) summit of Ben Hogh, with Tiree beyond

occupation, for it is cited as the base of Earl Gilli, brother-in-law of Sigurd, who ruled the Orkneys and the Hebrides around AD1000.

Measuring a modest twelve miles by three, Coll is formed of hard Lewissian Gneiss, which in the northern part of the island protrudes from the close-cropped turf like the ribs of a hungry cow. Curiously, the locals call the northern part of the island – between Arnabost and Sorisdale – the 'East End'. Arnabost to Hogh Bay is 'the North Side' and Arinagour to Breachacha 'the West End'.

As you head north, the landscape, like most of the island, is treeless, here hummocky, rugged and dotted with dark-eyed lochans. Across this, a single-track road teases a route with a commendable lack of directness, flowing unhurriedly onwards, twisting this way and that, poking into corners. At its end lies Sorisdale and a road (not for vehicles) leading down to a fine beach.

Along the North Side lie fine shell-sand beaches at Cliad Bay and Hogh Bay, along with the island's highest point, Ben Hogh, from which there is a fine view of the island and westwards to Barra and Uist. Near the summit of Ben Hogh a large boulder, according to folklore placed there by a giant, perches precariously on three smaller ones. At the road end, below the slopes of Ben Hogh lies a reed-fringed lochan, host to resident mute swans and numerous pairs of greylag geese that breed here. The passing breeze signs the surface of the loch, reeds bend, a coot coughs and the silence of pre-occupation settles once more on the feathered population.

Lamb on the roof of a ruined house at Scorisdale, Coll

fine example of Victorian negative improvement.

Beyond the road end and Loch Breachacha lies the vast, yawning void of Crossapol Bay, drawing the eye to the low land of Tiree. Due north, across a neck of grass-bound sand dunes, lies the equally empty curve of mezzaluna-shaped Feall Bay, and here, as at Hogh Bay and Cliad Bay, Atlantic rollers spend their force and crash relentlessly onto the shore.

At Totronald, concealed corncrakes croak and infuriate. Birdlife here exceeds 140 recorded species. The island in spring is awash with flowers, and among its waters otters, seals, pilot, minke and killer whales, dolphins, porpoises and basking sharks put in tantalising appearances, and then are gone.

TIREE

From the pitching ferry deck, approaching Tiree simmers with anticipation and warm, good feelings. A dark sliver of land rests lightly along the horizon, a thin line between sea and sky. Slowly the haze resolves: buildings appear at its edge, floating above the waves; rocks and low, foamedged headlands take form, and then the sandy miles of Tràigh Mhór, the 'Big Beach', across the gape of Gott Bay. It's the same every time: Tiree's

Left: Early purple orchid in a rocky landscape near Breachacha, Coll
Below: A gale whips up the sand on Tràigh Mhór, Gott Bay, Tiree

Down the West Side, the road leads to Breachacha Castle, built on a low promontory at the head of Loch Breachacha in the first half of the fifteenth century. This was home of the Macleans of Coll until 1750, when the thirteenth Chieftain, Hector Maclean, moved to a new and somewhat unattractive castle a hundred yards away, though parts of the old castle continued in use for many years. Samuel Johnson was not impressed with the new structure, commenting to Boswell that 'there was nothing becoming a Chief about it: it was a mere tradesman's box'. Doubtless he would have been further dismayed had he seen the parapet and pepperpot turrets added later, a

way of saying 'Welcome'. The boat berths and Tiree is underfoot.

Known to Gaelic speakers by two names – Tiriodh ('Tir', land, and 'iodh', corn) and Tiridhe ('Idhe', the genitive case of 'I', Iona) – Tiree is also sometimes referred to as Tir-fo-Thuinn, 'the land beneath the waves', and with just cause. Most of the island, this sunshine isle, is so low lying it is easy to fear that one good tsunami would erase it from the map altogether.

But what Tiree lacks in elevation, it compensates with rich machair and wide, golden miles of beach that each year host the International Windsurfing Championships.

The most westerly of the Inner Hebrides, Tiree is uncomplicated and relaxing, a place where the dazzling beauty of its white, shell-sand beaches onto which the warm waters of the Gulf Stream roll endlessly, has a mesmeric effect, willing you to spend time in their company and neutralising any incentive to move on. Cold temperatures are rare here, and the island holds Britain's sunshine record.

A scatter of houses at Scarinish lays claim to being the island's 'capital', though none is needed. At Sandaig, a row of restored white-houses present a museum of island life, and at Hynish an old signalling tower maintains its link with the distant Skerryvore light. As with all the island's communities, Kirkapol is here and there, a ragged, strung-out community drawn together by the thread of mutual dependence.

The first mention of the island is found in an ancient poem by Fionn Mac Rosa Ruaidh, who remarks on the destruction of eight towers in Tiree around 200BC. The island found Christianity long before St Columba reached

Iona, but it is from the *Life of St Columba* that we read the earliest history of Tiree, essentially religious in nature and clearly underscoring the belief that Tiree was indeed 'Tiridhe', the Garden of Iona, and very much under the rule of the monks there.

After the monks, in common with the rest of the Hebrides, Tiree was ruled by the Vikings, until

Opposite: Washing-line at Balemartine, looking over Hynish Bay. Tiree is reckoned to be the windiest place in Britain, though North Ronaldsay in Orkney makes a similar claim
Above: The Ringing Stone, Tiree

the defeat of King Haakon at the Battle of Largs in 1263. Three years later, the island was subsumed into the Lordship of the Isles, and experienced a chequered, and often brutal, ownership to reach the present hands of the Duke of Argyll.

Among its ancient monuments, the most renowned is the Ringing Stone, a large erratic brought by ancient glaciers from Rum. When struck with a small rock, it 'rings' almost musically, but is famed, too, for cup markings on its surface, thought to have been fashioned by the hand of prehistoric man.

From Balephetrish a track runs behind the farm and north-east along the shore towards Vaul. Clover, orchid, rattle and bedstraw brighten the machair; pimpernel, iris, forget-me-not and mint dot the reef marshlands. Along the tide-edge, a shelduck mum mollycoddles her twelve mottled young. Among the machair-boulders an oyster-catcher chick, rock-hued, places its faith in camouflage – a faith that would have been well-founded had not chance steered me to within inches of where it lay, rigid with fear. Out there, somewhere, a red-ringed eye watches from the shadow of the rocks: an anxious parent, apprehensively silent. I hastened by, elated, grateful for the encounter, watchful for siblings.

Left: Cul Sgathain, Tiree, looking to Hynish, with its signal tower and keepers' cottages built in the 1830s to support construction of, and communication with, the lighthouse on Skerryvore, about ten miles to the south-west

Opposite : Near Balinoe, looking towards Sandaig, Tiree

Later, I stood with my back against the pitted surface of the Ringing Stone, watching grey seals at the water's edge and lines of low-flying tysties skimming the waves. Further out, a bright-yellow fishing boat came and went with the swell, its rise and fall a lazy pulse against which life in the Scottish islands seems to be measured. Exposure to too much of this island magic, I thought, could mark a man for life, leaving him forever entangled in inexplicable enthusiasms for lazy days, bright blue skies, an encircling sea and the quiet ways of island nature. Tiree is the cradle of such feelings, birthplace of simple desires.

Left: Rock sculpture, Ceann a'Mhara, Tiree
Above: Snail, in the dunes, Gott Bay, Tiree

THE SLATE ISLANDS
SEIL, EASDALE, LUING

Lush and flower-rich Seil Island is tethered to the mainland by an imposing single-arch bridge designed by Robert Mylne and built by a Mr Stevenson of Oban, known properly as Clachan Bridge, but locally, for the past thirty years, as the 'Bridge over the Atlantic'. Among its mossy cracks grows the fairy foxglove (*Erinus alpinus*), a refugee, as its scientific name suggests, from the mountains of Europe.

As with the neighbouring islands of Easdale and Luing, Seil is renowned for its slate, and a number of small villages, built originally to house workers in the quarries, still remain, notably at Ellanbeich where the row of white cottages featured in the film adaptation of Gavin Maxwell's *Ring of Bright Water*.

Immediately adjoining the Clachan Bridge is a pub, the Tigh na Truish (the house of the trousers). There is a suggestion that the name derives from a time when the wearing of Highland dress was forbidden by law, and so it was here, before setting foot on the mainland, that islanders changed into a pair of breeks. This implausible notion overlooks the fact that the law applied equally throughout Scotland, not just to the mainland. More realistically, the house probably earned its name from being the residence, according to the 1791 census, of a family of trousermakers.

On the quayside at Easdale, Corrie and Rusty, two now-ageing dogs, await each ferry's

Above right: Clachan Bridge and Tigh na Truish, Seil
Right: Fishing boat at Cullipool on Luing, with Fladda and the Garvellachs behind

Above: Kerrera from the evening ferry from Mull to Oban
Opposite: Whirlpool in the Grey Dogs, a notorious tide-race between Scarba and Lunga, south of the Slate Islands

community in November 1881, when the combination of a storm and a high tide filled most of the quarries with water. Thankfully, no lives were lost (except that of a cat at Cullipool), and the quarries were soon able to resume producing slate, which they did for a further thirty years.

Luing is the largest of the Slate Islands, and it, too, once played a key role in the production of commercial slate, which was quarried here until 1965, close by the principal village of Cullipool.

KERRERA

Barely five minutes of bucking ferry ride on an open boat from Gallanach, the car-free island of Kerrera is a timeless world apart, one of green, hummocky hills where the only thing that moves on is the curious day-visitor.

But therein lies the island's charm.

There is a vaguely utilitarian feel to the island, at odds with its air of calm rurality. It was, and still is, a massive, natural breakwater sheltering the town of Oban. And since the eighteenth-century days of cattle droving, when Oban (according to Dr Johnson) was 'only a small village', Kerrera has been a stepping stone between Mull and the mainland. From Mull as many as 2,000 beasts a year would swim across the Firth of Lorn from Grass Point to Barr nam Boc, and then from the northernmost point of Kerrera to Dunollie. Today, these old drove roads serve as an ample, efficient and pleasurable way of exploring the island.

At the southern end of the island, turning its back on the clamour-echoes of Oban, and easily reached from a tea garden at Lower Gylen – and it is a tea 'garden', for there is only outdoor seating – stand the remains of Gylen Castle, still largely in the ransacked state left by General Leslie and the

visitors to guide them, free of charge, on the twenty-minute stroll around the island. By all accounts, they make a forlorn sight in winter when no one new arrives.

Just as saddening is the sight of so many quarrymen's cottages, now converted to holiday home use, though one does admirable duty as the island's folk museum. Disaster struck this tiny

Covenanting Army, who attacked it in 1647. The castle, now under restoration by Historic Scotland, was built of local stone in 1582 by Duncan MacDougall of Dunollie, on the site of an earlier fortification. By far the most impressive fortification in the Oban area, Gylen Castle, poised watchfully on a cliff overlooking the Firth of Lorn, commands a fine view of Seil, Lunga and Scarba.

Northward, in the area known locally as 'the Free State', busier and with the greater share of inhabitants, a memorial column to David Hutcheson, one of the founder members of the company that was to become Caledonian MacBrayne, stands on the point.

LISMORE

This long, fertile, flower-laden island in the jaws of Loch Linnhe is quiet, unspoiled and has a surprising knack of losing you in its folds. In *Gulfs of Blue Air*, Jim Crumley writes: 'If you have never been to Lismore at Orchid time, go before you die. Lismore has orchids like lesser islands have grass, or bogs, or rocks'.

The island is geologically odd, being composed almost entirely of Dalriadan limestone, which produces a fabulous display of flowers. More significantly, it is located across the Great Glen fault line, and minor earth tremors are a frequent occurrence.

The early history of Lismore revolves around St Moluag, a contemporary of St Columba who came to Scotland to spread the Gospel among the Picts. Legend records that both saints coveted the lush island and, as they 'currached' frantically for the shore, Moluag is said to have cut off his little finger and thrown it forward onto the island, so claiming it for himself.

In the thirteenth century, Lismore became the seat of the Bishop of Argyll, and the remains of the first cathedral are now incorporated into the tiny church at Clachan. The Bishop, however, chose to live at Archadun (Achinduin) Castle on the south-west coast of the island, overlooking the neighbouring nearly tidal island of Bernera.

On the eastern shore at Balure is a galleried broch, one of the best preserved monuments in Scotland, known as Tirefour Castle, and thought to date from the time of Christ. The people who built it, however, were not the first inhabitants of the island, for a polished Appin stone axe-head, dated to 3500BC, was found at Balnagown. On the north-west coast stand the ruins of Castle Chaifen or Coeffin. This appears to have been built by the MacDougalls of Lorn on the site of a Viking fortification.

Bernera Island, accessible on foot, usually, at low tide, was once known as Berneray of the Noble Yew, an allusion to a yew tree that even in Moluag's day had so great a spread that a thousand people could shelter beneath its branches. Alas, such pedigree is nought compared to the need of Lochnell Castle for a staircase; the tree was felled about 1850.

Opposite: Waterfall at Port nan Urrachann, Scarba
Right: Lismore parish church

FIRTH OF LORNE, KINTYRE AND THE CLYDE ISLANDS

ISLAY AND JURA

These two islands lie close together but are remarkably different in character. Islay (pronounced Eye-la) is for the most part low-lying, with moorland and a few small hills in the southeast buttressed by some fine sea cliffs. Jura is almost the reverse, being rugged and hilly, an outstanding island wilderness from the centre of which the distinctive Paps of Jura gaze out on the middle of nowhere and further afield. The two are separated by the narrow Sound of Islay, less than 800 yards wide at its narrowest. Here a ferry plies between Port Askaig and Feolin Ferry, and this is as good a place as any at sea level to point out the differences between Islay and Jura.

Islay is renowned for whisky and wildlife – both of which occur in abundance – rich

Opposite: A winter evening at Machir Bay, Islay

Right: Bunnahabhainn distillery, Islay, from a ferry in the Sound of Islay

Above: Heron, Loch Indaal and the Paps of Jura from
Bowmore Pier, Islay
Opposite: Jura from Port Askaig on Islay, with early clouds
clearing from the Paps

agricultural land and floral diversity, and has as
much in common with Ireland as with Scotland
being, at Mull of Oa, almost equidistant between
the two. Between the sixth and ninth centuries,
Islay was part of the Kingdom of Dalriada and

ruled from the northern part of Ireland.

The island used to be the seat of the
Lordship of the Isles in the fourteenth and
fifteenth centuries, the most important of all the
Scottish islands, and a power-base sufficiently

strong to enter into its own treaties with England, Ireland and France, and not always in Scotland's best interest. The headquarters of the lords was at Loch Finlaggan, north of Ballygrant, and islands in the loch support crumbling ruins including a great hall and a chapel. Forlorn now, they hardly convey the impression of the power they once housed.

Standing stones, crannogs, duns, forts and chambered cairns proliferate, and tell of Islay's prehistory. But few of these relics compare with the Kildalton High Cross, close by Ardmore

Point. Carved by a sculptor from Iona around AD800, the cross ranks among the finest Celtic relics in Scotland.

Loch Gruinart, Loch Gorm and Loch Indaal offer shelter to visiting wildfowl; over-wintering Barnacle geese and Greenland white-fronted geese appear in great number, and make a memorable sight as they come in to roost. Among the cliffs at the Mull of Oa, wind-slipping choughs ride the air currents, sharing the territory in winter with snow bunting, purple sandpiper and a colourful herd of feral goats.

Above left: Still room, Ardbeg Distillery, Islay

Above: Interior of the eighteenth-century round church at Bowmore, Islay

Right: Port Charlotte, Islay, a 'model' village built around 1829

Above left: Kildalton High Cross, Islay

Above: Eilean Mhor, Loch Finlaggan, Islay: ruins of the Great Hall, looking towards the crannog of Eilean na Comhairle (Council Island)

Left: Red deer stag, near Tarbert, Jura

Opposite: Barnacle geese at the head of Loch Indaal, Islay, with Bowmore in the background

Opposite: Loch na Mile and Craighouse from Maol nam Fineag, Jura

Above: Even on a calm day, mysterious waves and whirlpools disturb the waters of the notorious Gulf of Corryvreckan. The north end of Jura is seen behind

Underlying metamorphic rock produces a barren and dour moorland landscape on Jura, devoted almost entirely to deer farming, though there seem to be almost as many adders as there are deer – if you know where to look. Jura is by far the wildest island among the Inner Hebrides. Much of the island is delightfully roadless and uninhabited with vast areas of blanket bog, heather, bracken and rock, that culminate in the three Paps of Jura, summits almost as high as

those on Arran, and certainly as widely famed as any in Scotland.

Only one road penetrates this wilderness, though 'penetrate' is hardly the word for a single-track ambition that follows the east coast. Beyond Lealt the road degenerates to a vehicle track and leads past Barnhill, home of Eric Blair (aka George Orwell) who lived here while writing his prophetic novel *1984*. From the northernmost point of the track a path continues to the cliffs at Carraigh Mhor, overlooking the notorious Corryvreckan whirlpool, a natural phenomenon at its best when a strong west wind meets a spring tide in flood.

COLONSAY AND ORONSAY

Linked by a tidal strand, Colonsay and Oronsay rival Tiree and Sanday (Orkney) for golden beaches. Halfway across The Strand is the sanctuary cross, which it is said, if reached by any Colonsay villain, gave immunity from punishment provided he (and presumably, she) remained on Oronsay for a year and a day. Local legend has it that it was Oronsay that was first visited by St Columba on his exile from Ireland, but on ascending Beinn Oronsay's modest elevation he discovered that he could still see Ireland, and so sailed on.

Both islands have been occupied since at least 5000BC: Oronsay has a number of Mesolithic shell middens, and thirteen duns and eight forts, some dating from 500BC, are dotted around the coastline. Augustinian Canons occupied the magnificent fourteenth-century priory on Oronsay, which is thought to stand on the site of an earlier monastery dating from the sixth century. The priory became an important

Opposite: Rainbow over Jura, from the Islay ferry
Above: Riasg Buidhe Cross, in the grounds of Colonsay House

Above left: Palm trees in the woodland garden near Colonsay House
Left: Oronsay from the evening ferry to Port Askaig
Above: Atlantic breakers, Kiloran Bay, Colonsay
Opposite: Kiloran Bay

religious centre for Argyll and its islands over the ensuing 200 years. The ruins are well preserved, and what is missing may well be found disguised as nearby farmhouses.

The islands have a convoluted coastline running, along the west coast, to machair and raised beaches that tell of a time when two may well have been four. The isolation makes Colonsay and Oronsay ideal for visiting birdlife, which, though lacking such seasonal rarities as may be found elsewhere among the Scottish islands, nevertheless yields a commendable tally.

GIGHA

Barely six miles by two, Gigha is famed for the gardens at Achamore House, created by Sir James Horlick, creator, too, of a night-time drink. But rightly famed as they are, they are no match for Nature's display of flowers from bluebells to the Grass of Parnassus; in June the flanks of the island's five-mile road are intensely awash with colour and alive with birdsong. It's as if Nature has found its own Paradise, and doesn't mind who knows.

On a neck of land separating East and West Tarbert Bay, the so-called Druid Stone, a tilted monolith of modest proportions guides the eye to the hills of Jura – as if any guiding was needed – and demonstrates better than anything, the qualities needed for island exploration – patience, studied observation, reflection, lack of haste and a profound degree of nonchalance. The ingredients – time and money – are an investment that knows nothing of stock markets and a good deal about rewards.

Beside the eloquent ruins of St Catan's Chapel at Kilchattan is an Ogam stone, the only one of its kind in the west of Scotland, and mellow-soft as the sun slips out of sight beyond the darkening bulk of Islay and Jura. Cinnamon-scented, razor-sharp dwarf whin crowds the island's summit, the best place to be when the sun goes down.

Opposite: Achamore Gardens, Gigha: the Malcolm Allen Garden, commemorating a former head gardener who served for fifty-two years

Right: Highland calf at the north end of Gigha, with Jura on the horizon

ARRAN AND HOLY ISLE

By the roadside at Machrie Bay a pair of oystercatchers confront a ewe that has wandered perilously close to where their eggs lie concealed among pebbles. Neb and nose close to a matter of inches, as if engaged in conspiratorial bargaining. Then, negotiations in tatters and no deal struck, one of the birds leaps into the air and bounces repeatedly on the ewe's back – there are some for whom the word of reason simply does not function, it seems to be saying. It is a featherweight protest in the scale of things, but sufficient for the ewe to get the message and send her on her way. Oystercatchers One, Sheep Nil.

From afar, Arran doesn't seem like an island at all. Visitors, heading overland to Ardrossan, see only a range of hills rising beyond the landscape of Ayrshire. And seen thus, Arran melds easily into the backdrop of Kintyre, indistinguishable from it. Only as the visitor draws close is the Firth of Clyde revealed, and only from the breezy deck of the Brodick-bound ferry does the great distinction between the craggy outburst of the north and the fertile farmlands of the south finally come into view. Between the two lies that great geological signature, the Highland Boundary Fault, but you would need to be perched high on the slopes of Goat Fell to see the distinction at its best. Little surprise then that Robert Burns, blinkered with the minutiae of his Ayrshire homeland, failed to recognise what has inspired artists and poets a-many.

For generations of Glaswegians, Arran was but a trip 'Doon the watter', and part of the county of Buteshire. That it continues to shoulder a disproportionate burden of tourism is testimony to the island's capacity to provide something for everyone. Measuring nineteen miles by ten, with an area of 165 square miles, Arran contains many of the features traditionally associated with both Highland and Lowland scenery; its position astride the faultline sees to that. As a result, it is fashionable, if not a little fanciful, to speak of Arran as 'Scotland in miniature'.

Although portrayed as one of the 'Clyde Islands', Arran's characteristics more truly reflect those of the Inner Hebrides – geologically, culturally and historically. Much of the superficial landscape was fashioned following the Clearances in the eighteenth and early nineteenth centuries, and there is a definite Celtic influence that underpins ancient links with Ireland and the legendary Irish king, Fionn McCool. These qualities are not readily perceived from a quick tour of the island, but come to mind when it is explored leisurely and on foot. For then it is that the moors disclose their stone circles, monoliths and chambered cairns, the straths unveil the remains of ancient chapels, black-houses and turf dykes, the hilltops reveal the traces of fortifications, and the language of Norse speaks aloud in placenames and features across the landscape: Brodick derives from Broad Vik, Sannox from Sandy Vik, and the name of Arran's highest peak, Goat Fell, from Geita Fjell.

Its mountains are by far Arran's most dominant features, and genuinely appreciated by

the hill-walking fraternity for they offer a splendid group of jagged peaks to rival, visually at least, those of Skye. Closer acquaintance, however, reveals a more friendly disposition towards walkers with less demanding ambitions, though there are countless nooks and crannies among Sheriff Nicolson's 'terrible congregation of jagged mountain ridges and fantastic peaks' to test the most resolute of adventurers.

The first of these visitors, facing even greater tribulations, would have arrived on Arran in small family groups, travelling in dugout canoes and skin boats, having crossed the sea from Northern Ireland or travelled up the coast from Galloway. With little to lure them inland, where the land was still too poor to support the production of crops, these early settlers set up their skin tents around the coasts, living on a diet of roots, berries, fish and shellfish. Pollen analysis from around Lochranza suggests these Mesolithic visitors came about 6500BC: had it been eight and a half millennia later, they could have celebrated their arrival with a dram of Scotland's newest malt whisky.

About 4,000 years BC, Neolithic man arrived on Arran, bringing 'new' tools and a way of life based on agriculture. But for a flavour of such distant times, visitors must move forward to the Bronze Age and the fabulous arrangement of stone circles found across the southern part of the island, especially on Machrie Moor. Here, in a lowland area overlooked by smooth-flanked uplands and the granite peaks of the north, is one of the most remarkable archaeological sites in Scotland, comprising the ruins of chambered tombs, hut circles and six megalithic rings. Their variety suggests they were constructed over several centuries, and traces of hut walls and enclosures further west imply that they stand in what may have been a sacred area, where 'dwellings' were forbidden.

Close by, along the shoreline at Drumadoon, are a series of sea-fashioned caves above a raised beach, known today as the King's Cave. Here, it is said, the future king of Scotland, Robert Bruce, was inspired by the persistence of a spider building a web to return to his quest for the throne of Scotland. The attribution of the King's Cave to Robert Bruce is dismissed by historians, though this should not deter visitors from the walk to this splendid site. Even if the cave is barred by massive iron gates (to prevent further despoliation by idiot graffiti-mongers), and the carvings on the cave walls – thought to be authentic murals of the Scots of Dalriada – also overgrown with moss, there will be no disappointment in what is found here.

Bruce, however, was not the first of Arran's royal visitors. In 1263, King Haakon IV of Norway anchored his fleet in Lamlash Bay en route to meet the forces of Alexander III at the Battle of Largs. King Haakon was defeated, and this removed the long-standing threat to Scotland. The king died in Kirkwall in Orkney on the way home, and three years later his successor ceded all the western isles to the King of Scots.

Today's sea-borne visitors arrive at Brodick, overlooked by the magnificent Brodick Castle – ancient seat of the Dukes of Hamilton and more recently the home of the Duchess of Montrose – and shapely Goat Fell, both now in the care of

Opposite: Winter on Arran's peaks: Cir Mhor and Caisteal Abhail from Goat Fell

Right: Standing stones, Machrie Moor, Arran

the National Trust for Scotland. Further south, however, Lamlash Bay, where King Haakon sheltered before pressing on to Largs, is protected by shapely Holy Island. The trip across the bay to the island is today a popular one, supported by a ferry service operating from a caravan on the shore at Lamlash.

The central Mullach Mor gives Holy Island an impressive appearance, and inevitably draws the curious across. Known in the Celtic language as Innis Shroin, the 'island of the water spirits', Holy Island has a long history as a place of spiritual importance. The early Christian saint, Molaise (Mo Las), settled on Innis Shroin in the sixth century having rejected the crown of Ulster to which he was entitled. The cave in which the saint lived is still a place of pilgrimage, as is the Judgement Rock, where he is said to have sat while giving advice to those who sought the

benefit of his wisdom. Nearby, complete with drinking utensil, is a holy well whose clear waters are said to have healing properties.

Now under the stewardship of the Samye Ling Buddhist Community, the island has become a focus for work on three great concerns of our time: the environment, peace and spirituality. Holy Island still remains relatively unharmed, and whatever is done on the island will respect the environment.

Less esoterically, Holy Island is indeed a place of quiet reflection. Buddhist carvings and rock paintings decorate the island, bringing a bright if bizarre touch of surrealism to the place. The island has a rich ecological heritage that embraces many rare plants, a flock of Soay sheep, a growing population of feral Sanaan goats and a small herd of the rare Eriskay ponies.

Left: Common seal, Brodick Bay
Above: Yachts and Brodick Castle, Arran
Opposite: Lamlash golf course, Arran, with Holy Island beyond

GREAT AND LITTLE CUMBRAE ISLANDS

Carved from sandstone boulders, two figures stand watch over visitors to Great Cumbrae's shores. The figures are historical and the sculpting, known as 'Cumbrae Passage', is intended to 'explore the intensity of the moment of arriving or leaving'. Neither figure belonged to the island, but both came to play a pivotal part in its story.

Today, Great Cumbrae, within easy reach of the Scottish mainland, is the most densely populated of the Scottish islands, and has been a popular outing since the heyday of the Clyde steamers, which experienced a mild hiccough in 1906 when they boycotted the island over the issue of harbour fees. It was a conflict resolved only by the timely intervention of Lloyd George, then President of the Board of Trade in Campbell-Bannerman's government.

Millport, 'capital' of the island, is a typical seaside resort without the brashness, and boasts the smallest cathedral in Britain, begun in the 1840s by the 6th Earl of Glasgow. At the top of the island, Tomont End is where King Haakon of Norway set up camp as he surveyed the debacle of the Battle of Largs. Apart from the 'urbanism' of Millport, the island enjoys a rippling coastline of sea-shaped old red sandstone rising to a green ridge of hills crowned by the Glaid Stone and a hill road with unrivalled views of Arran.

Little Cumbrae is much less accessible, a rugged place with a coastline of rock outcrops,

Opposite: Eroded rocks at Scalpsie Bay on Great Cumbrae, looking to Little Cumbrae and Arran
Right: The Cathedral of the Isles, Millport, Great Cumbrae

caves and a raised beach at the southern end. On the highest point of the island stands a lighthouse, constructed in 1757, and of the type known as a cresset that had a fire of coals burning in an open grate. Pity the poor men who laboured to ferry coal up the hillside and onto the fire platform, and then struggled to keep the fire burning in all weathers.

Above: Scalpsie Bay, Bute, looking to Arran
Opposite: Millport, Great Cumbrae

BUTE

A promenade beetling with hotels, guests houses and B&Bs, pastel-painted porticoes, curlicues, columns, cast-iron coronets, flutings and sundry other Victorian excesses greets your arrival at Rothesay. Glasgow tongues 'frae up the watter' leaden the air and skirmish with the alien sounds of incomers from further afield. Jostling for service in the award-winning chippie or queuing to see the restored Victorian lavatories on the pier, there's no mistaking the holiday atmosphere – frenetic, flippant and fulsome.

A few hundred yards from the shore, moated Rothesay Castle, favoured residence of kings Robert II and III and of strategic importance to both James IV and V during their Hebridean campaigns, formerly marked the shoreline, the intervening ground being made up largely during the seventeenth and eighteenth centuries.

Thankfully, there is a more substantial island beyond, a serene place of sandy beaches, promontory forts, Celtic chapels, stone circles, standing stones, flowery banks and hedgerows, gentle hills and lush green farmland. In the centre of the island, freshwater Loch Fad and Loch Quien betray the Highland Boundary Fault: to the north the land, based on metamorphic rocks, is largely wild infertile moor, hilly, uninhabited; to the south such a contrast exists where old red sandstone feeds a pastoral landscape resembling Ayrshire at its best.

Opposite: Rothesay harbour

Above: Victorian lavatories, Rothesay, Bute

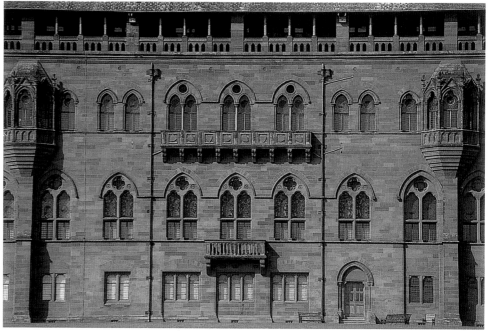

Left: Victorian villas, Rothesay, from the ferry
Below: The east frontage of Mount Stuart, Bute
Opposite: Holy Island from Lamlash, Arran

Most of the island is owned by the Marquess of Bute, whose family have held sway over the island for over 250 years, ruling, since its construction in the nineteenth century, from the great Gothic palace of Mount Stuart, built by the 3rd Marquess of Bute in an attempt to revive Medieval architecture, and incorporating elements that are as much European as British.

At the southern tip, sheltered by a veil of ash and elm, the infinitely more humble but atmospherically far better endowed twelfth-century St Blane's chapel is on the site of an early monastery and at the centre of a Christian complex which pre-dates it by up to 600 years. It is a gentle and serene spot, given to quiet and peace.

Nearby, the ruins of a fort overlooking Dunagoil Bay serve both to remind visitors of distant times and to provide a perfect balcony seat from which to survey the ruggedness of adjacent Arran.

BIBLIOGRAPHY

So many books have been written about the Scottish islands over the years. All are useful and informative, the vast majority well written. But for anyone suitably inspired by the present work to set out to explore these islands for themselves, there are three authors whose work stands above all others and will help in an understanding.

The indispensable 'bible' to the Scottish islands is Hamish Haswell Smith's *The Scottish Islands* (though I actually gave mine away to the monks on Papa Stronsay, and had to buy another when I got home). For the finest nature writing, read Jim Crumley; for descriptions, observations and explanations of the most intimate detail of island life, read Derek Cooper's excellent works.

Argyll: An inventory of the monuments: Volume 3, Mull, Tiree, Coll and Northern Argyll (The Royal Commission on the Ancient and Historical Monuments of Scotland, 1980)

Ashmore, Patrick. *Maes Howe* (Historic Scotland, 2000)

Booth, David and Perrott, David. *The Shell Book of the Islands of Britain* (Guideway/Windward, 1981)

Boswell, James. *The Journal of a Tour to the Hebrides with Samuel Johnson LLD* (J.M. Dent & Co, 3rd ed 1909; Everyman's Library, ed Ernest Rhys)

Bray, Elizabeth. *The Discovery of the Hebrides: Voyages to the Western Isles, 1745–1883* (Birlinn Ltd, 1986, 1996)

Brownlee, Niall M. *Townships and Echoes* (Argyll Publishing, 1995)

Clarke, David. *Skara Brae* (Historic Scotland, 1989, 1996)

Cluness, A.T. (ed). *The Shetland Book* (Zetland Education Committee, 1967)

Cooper, Derek.
– *Skye* (Birlinn Ltd, 1995)
– *Hebridean Connection: A View of the Highlands and Islands* (Routledge and Kegan Paul, 1977)
– *The Road to Mingulay: A View of the Western Isles* (Futura Publications, 1988, Warner Books, 1992, 1999)

Craig, David. *On the Crofters' Trail* (Jonathan Cape, 1990–1992)

Crumley, Jim.
– *Among Islands* (Mainstream Publishing, 1994)
– *Gulfs of Blue Air* (Mainstream Publishing, 1997)
– *The Heart of Mull* (Colin Baxter, 1996)
– (with Colin Baxter) *Shetland: Land of the Ocean* (Colin Baxter, 1995, 1998)
– (with Colin Baxter) *St Kilda: A Portrait of Britain's remotest island landscape* (Colin Baxter, 1988)

Fisher, Andrew. *A Traveller's History of Scotland* (Windrush Press, 1990, 1997)

Fojut, Noel. *The Brochs of Gurness and Midhowe* (Historic Scotland, 1993, 1996)

Gemmell, Alastair. *Discovering Arran* (John Donald Publishers Ltd, Edinburgh, 1990, 1998)

Gordon, Seton. *Highland Days* (Cassell and Company, 1963, Country Book Club, 1965)

Grant, James Shaw. *Discovering Lewis and Harris* (John Donald, Edinburgh, 1987)

Grimble, Ian. *Scottish Islands* (BBC, 1985)

Haldane, A. R. B. *The Drove Roads of Scotland* (House of Lochar, 1995)

Haswell-Smith, Hamish. *The Scottish Islands* (Canongate Publishing, 1996–2001)

Hedderwick, Mairi. *An Eye on the Hebrides: An Illustrated Journey* (Canongate Publishing, Edinburgh, 1989)

Howard, J. and Jones, A. *The Isle of Ulva: A Visitor's Guide* (self-published, 1990)

Hunter, James. *Last of the free: A Millennial History of the Highlands and Islands of Scotland* (Mainstream, 1999)

Kennedy, Donneil. *The land below the waves: Tiree past and present* (Tiree Publishing Company, 1994)

Knapman, Meena J. A., *Discover the Isle of Tiree* (Tiree Community Business, 1997)

Linklater, Eric.
– *Orkney and Shetland* (Robert Hale, 1965)
– *The Prince in the Heather* (Hodder and Stoughton, 1965–1966)

Lismore Historical Society. *Lismore* (1995)

MacArthur, E. Mairi. *Columba's Island: Iona from past to present* (Edinburgh University Press, 1995, 1996)

MacDougall, Betty. *Guide to Coll* (1986)

MacGregor, Alasdair Alpin.
– *An Island here and there* (Kingsmead, 1972)
– *Islands by the Score* (Michael Joseph, 1971)
– *Over the Sea to Skye, or Ramblings in an Elfin Isle* (W. & R. Chambers Ltd, 1930)
– *The Western Isles* (Robert Hale, 1949)

Maclean, Charles. *St Kilda: Island on the Edge of the World* (Canongate, 1972–1996)

Magnusson, Magnus. *Rum: Nature's Island* (Luath Press, 1997)

Martin, Martin. *A Description of the Western Islands of Scotland Circa 1695* (Birlinn, Edinburgh, 1999)

Marwick, Hugh. *Ancient Monuments in Orkney* (HMSO, 1952)

McCrum, Mark. *Castaway* (Ebury Press, 2000)

McLellan, Robert. *The Isle of Arran* (David & Charles, 2nd ed, 1976)

Miller, Ronald. *Orkney* (Batsford, 1976)

Mitchell, W.R.
– *It's a Long Way to Muckle Flugga: Journeys in Northern Scotland.* (Souvenir Press, 1990)
– *St Kilda: A Voyage to the Edge of the World* (House of Lochar, 1999)

Newton, Norman. *Skye* (Pevensey Island Guides, 1995)

Orr, Willie. *Discovering Argyll, Mull and Iona* (John Donald, Edinburgh, 1990)

Ritchie, Anna. *Orkney* (The Stationery Office, 1996)

Schei, Liv Kjørsvik. *The Islands of Orkney* (Colin Baxter, 2000)

Scottish Natural Heritage. *Arran and the Clyde Islands* (1997)

Thompson, Francis. *Harris and Lewis* (David & Charles, 1968).

Wickham-Jones, Caroline. *Orkney: A Historical Guide* (Birlinn, 1998)

Withall, Mary. *Easdale Island Folk Museum* (1997)

ACKNOWLEDGEMENTS

I have received immense kindness and help from everyone I met during my travels for this book; after years exploring in Scotland, I know such generosity is the norm, but it's not something to take for granted.

In addition, I have received help from many organisations, to arrange accommodation, organise and implement trips and provide valuable information. I record my indebtedness to them here: for assistance with travel – Caledonian Macbrayne Ferries; P&O Scottish Ferries; Loganair, Tingwall, Shetland; Virgin Trains. For help in organising trips and accommodation – Argyll, Loch Lomond, Stirling, the Trossachs and the Isles Tourist Board; Ayrshire and Arran Tourist Board; Orkney Tourist Board; Raasay Outdoor Centre; The Lerwick Hotel, Lerwick, Shetland; The Western Isles Hotel, Tobermory, Mull; Gruline Home Farm Guest House, Gruline, Mull; Garden House, Coll; Kirkapol Guest House, Tiree; Strathwillan House, Brodick, Arran; South Whillieburn Farm, Largs; Bayview Hotel, Rothesay, Bute. For invaluable help with information – The Scottish Slate Islands Heritage Trust; Ian Tait at Shetland Islands Museum, Lerwick.

Most important of all, I want to record my indebtedness to my wife, Vivienne, who spent her honeymoon exploring Orkney with me, organised me on all the trips, and endured the perils of rubber dinghies, pitching boats, tide races and my *al fresco* cooking with unflappable good grace.

Terry Marsh

For help with travel, I must particularly thank Caledonian MacBrayne. Their services are part of the fabric of life in the Clyde and the Hebrides, and now part of the fabric of this enterprise too. I am also very grateful to Loganair staff at head office and in both Orkney and Shetland. Generous help was also received from Arisiaig Marine; Turus Mara at Ulva Ferry, Mull; Sea.fari at Easdale, near Oban; and Sea Life Surveys at Tobermory, Mull. Thanks too to Serco Ltd and the National Trust for Scotland for making it possible for me to get to St Kilda.

For assistance with accommodation, I must, like Terry, thank the following: Argyll, the Isles, Loch Lomond, Stirling, and the Trossachs Tourist Board; Ayrshire and Arran Tourist Board; Orkney Tourist Board; and Raasay Outdoor Centre, who also gave me my only trip under sail during this period. The other hotels, guest houses and B&Bs I stayed in are too numerous to list but I would like to thank the Taylors at Leraback on Foula, who came up trumps with beds and meals at very short notice when my return flight evaporated into the mist.

I would also like to acknowledge the local communities and local authorities, especially in the Western Isles, who allow and facilitate free wild camping at some of the most inspiring locations in Britain. Long may this privilege continue: I can only hope it is never abused.

Finally, I must thank my parents, for keeping the rest of my business going while I was off enjoying myself. And above all, I must thank my partner, Bernie Carter, who submitted to requests to pose in some unlikely spots, and shared a rope on Sron na Ciche. Her own career did not allow her to accompany me on most of my other trips, and all those phone calls from far and wide must have been galling at times. For her unstinting support, I am more grateful than I can say.

NOTES ON PHOTOGRAPHY

The main landscape shots for this book were taken on 6 x 7cm format transparency film, using a Mamiya 7 camera with 43mm and 80mm lens. This camera uses rangefinder focusing and is extremely light and easy to carry. A Manfrotto carbon-fibre tripod was used for most of the shots with this camera.

Other material, including wildlife, action, long-lens and foul-weather shots, was taken on 35mm transparency film using Nikon cameras (FM2, F90X and F100) with lenses from 20mm to 600mm. 1- and 2-stop grey graduated filters were used sparingly, and polarising and 81B warm-up even more sparingly.

The film stock for the vast majority of shots was Fuji Velvia, with a few rolls taken on Provia 100. I briefly experimented with Kodak Ektachrome 100SW also.

Most interior shots were taken by available light but flash was necessary in a very few cases, such as some of the chambered tombs in Orkney where there is no natural light at all.

None of these images have been digitally manipulated beyond normal pre-press treatment such as the removal of dust and scratches. No changes whatsoever have been made to substantive picture content.

Jon Sparks

INDEX